Critical Acclaim for *The Mobile Connection*!

This powerful book illustrates the dramatic changes that have been provided by the social dynamics of the cell telephone and the ways that many long-held customs are changing: What is polite? How important is the time for a meeting when participants reschedule continually? What do we mean by a community or social group? And why are those short, inconvenient-to-type text messages more common and more important than voice conversations? Rich Ling provides a compelling examination of the real impact of mobile telephony. It's not about technology, it's about people. We need more of these kinds of studies.

—**Don Norman**, Co-Founder, Nielsen Norman Group and author of *Emotional Design*

For perhaps the first time in history, it is possible to gain scientific insights into the social impact of a new communication medium in the medium's infancy. Rich Ling combines scientific rigor, penetrating insight, and attention to an extraordinarily timely subject—the social impact of mobile communications. His ideas about "micro-coordination" and "the softening of time" are fundamental. Ling has big ideas about what the new world of always-on and ubiquitous media mean to our daily lives, but he's not an armchair theorist—he was smart and fortunate enough to observe the earliest adopters of mobile telephones first-hand.

—**Howard Rheingold**, Rheingold Associates and author of *The Virtual Community*

Rich Ling probes the way the mobile phone influences lives, talk and interaction. His carefully documented investigations paint an authoritative picture that will command continuing interest . . . an impressive achievement.

—**James E. Katz**, Ph.D., Professor of Communication, Rutgers University

We're shifting from wired to wireless. People are cutting loose from bounded groups, and want to connect anywhere and at anytime to their social networks. Cell phones lubricate this mobile-ized society. Rich Ling's pioneering work nicely pulls together the dance between mobile communication and the networked society.

—**Barry Wellman**, Ph.D., Professor of Sociology, University of Toronto

THE MOBILE CONNECTION

The Cell Phone's Impact on Society

The Morgan Kaufmann Series in Interactive Technologies

Series Editors: Stuart Card, PARC; Jonathan Grudin, Microsoft; Jakob Nielsen, Nielsen Norman Group

The Mobile Connection: The Rise of a Wireless Communications Society
Richard Ling

Information Visualization: Perception for Design, 2nd Edition
Colin Ware

Interaction Design for Complex Problem Solving: Developing Useful and Usable Software
Barbara Mirel

The Craft of Information Visualization: Readings and Reflections
Written and edited by Ben Bederson and Ben Shneiderman

HCI Models, Theories, and Frameworks: Towards a Multidisciplinary Science
Edited by John M. Carroll

Web Bloopers: 60 Common Web Design Mistakes, and How to Avoid Them
Jeff Johnson

Observing the User Experience: A Practitioner's Guide to User Research
Mike Kuniavsky

Paper Prototyping: The Fast and Easy Way to Design and Refine User Interfaces
Carolyn Snyder

Persuasive Technology: Using Computers to Change What We Think and Do
B. J. Fogg

Coordinating User Interfaces for Consistency
Edited by Jakob Nielsen

Usability for the Web: Designing Web Sites That Work
Tom Brinck, Darren Gergle, and Scott D.Wood

Usability Engineering: Scenario-Based Development of Human–Computer Interaction
Mary Beth Rosson and John M. Carroll

Your Wish Is My Command: Programming by Example
Edited by Henry Lieberman

GUI Bloopers: Don'ts and Dos for Software Developers and Web Designers
Jeff Johnson

Information Visualization: Perception for Design
Colin Ware

Robots for Kids: Exploring New Technologies for Learning
Edited by Allison Druin and James Hendler

Information Appliances and Beyond: Interaction Design for Consumer Products
Edited by Eric Bergman

Readings in Information Visualization: Using Vision to Think
Written and edited by Stuart K. Card, Jock D. Mackinlay, and Ben Shneiderman

The Design of Children's Technology
Edited by Allison Druin

Web Site Usability: A Designer's Guide
Jared M. Spool, Tara Scanlon, Will Schroeder, Carolyn Snyder, and Terri DeAngelo

The Usability Engineering Lifecycle: A Practitioner's Handbook for User Interface Design
Deborah J. Mayhew

Contextual Design: Defining Customer-Centered Systems
Hugh Beyer and Karen Holtzblatt

Human-Computer Interface Design: Success Stories, Emerging Methods, and Real World Context
Edited by Marianne Rudisill, Clayton Lewis, Peter P. Polson, and Timothy D. McKay

THE MOBILE CONNECTION

The Cell Phone's Impact on Society

Rich Ling
Telenor R & D

ELSEVIER

AMSTERDAM • BOSTON • HEIDELBERG • LONDON
NEW YORK • OXFORD • PARIS • SAN DIEGO
SAN FRANCISCO • SINGAPORE • SYDNEY • TOKYO

Morgan Kaufmann is an imprint of Elsevier

MK®
MORGAN KAUFMANN PUBLISHERS

Publishing Director	Diane D. Cerra
Publishing Services Manager	Simon Crump
Editorial Coordinator	Mona Buehler
Production Editor	Troy Lilly
Cover Design	Frances Baca Design
Cover Image	Getty Images, The Stone Collection
Photographer	Jonathan Storey
Composition	Cepha Imaging Private Limited
Copyeditor	Simon and Associates
Proofreader	David Weintraub
Indexer	Joy Dean Lee
Interior Printer	The Maple-Vail Book Manufacturing Group
Cover Printer	Phoenix Color Corporation

Morgan Kaufmann Publishers is an Imprint of Elsevier.
500 Sansome Street, Suite 400, San Francisco, CA 94111

This book is printed on acid-free paper.

Library of Congress Cataloging-in-Publication Data

Ling, Richard.
The mobile connection : the cell phone's impact on society / Richard Ling.
p. cm.
Includes bibliographical references and index.
ISBN: 1-55860-936-9 (pbk. : alk. paper)
1. Cellular telephone systems—Social aspects. I. Title.
HE9713.L563 2004
303.48′33—dc22

2003028268

ISBN: 1-55860-936-9

For information on all Morgan Kaufmann publications,
visit our website at *www.mkp.com*

Printed in the United States of America
05 06 07 08 5 4 3 2

To Grandmommy Seyler for her sense of curiosity,
and to Grandma Ling for her good common sense.

Contents

Preface xi

CHAPTER 1

Introduction 1

Introduction 1
History of Mobile Telephony 6
Growth of the Mobile Market 11
Outline of the Book 17

CHAPTER 2

Making Sense of Mobile Telephone Adoption 21

Interaction Between Technology and Society 21
Technical/Social Determinism and Affordances 23
Domestication of Information and Communication Technologies (ICTs) 26
Methods and Data Sources 33

CHAPTER 3

Safety and Security 35

Introduction 35
The Mobile Telephone as a Contribution to Security 37
Use in Situations Where There Is a Chronic or Acute Need for Contact 38
Abstraction of Security vis-à-vis Mobile Telephony 42
The Mobile Telephone in Extraordinary Situations 46
Diminution of Safety 43
Driving and Mobile Telephone Use 49
Personal Privacy 53
Conclusion 54

CHAPTER 4

The Coordination of Everyday Life 57

Introduction 57
Social Coordination 61
Mechanical Timekeeping and Social Coordination 63
The Development of Mechanical Timekeeping 64
The Standardization of Time 65
The Etiquette of Time and Timekeeping 67
Mobile Communication and Microcoordination 69
Midcourse Adjustment 70
Iterative Coordination 72
Softening of Schedules 73
Time-Based vs. Mobile-Based Coordination 76
Advantages of Mobile Based Coordination 76
Limitations 77
Competition or Supplement? 78
Conclusion 80

CHAPTER 5

The Mobile Telephone and Teens 83

Introduction 83
Child/Adolescent Development and the Adoption of Telephony 86
Adolescence and Emancipation in Contemporary Society 93
Elements in the Adoption of Mobile Telephony by Teens 97
Functional Uses of the Mobile Telephone Among Adolescents 99
Symbolic Meaning of the Mobile Telephone 103
Social Networking via the Mobile Telephone 110
Monetary Dimensions to Teens' Adoption of Mobile Telephony 112
Conclusion: Mobile Telephony and the Dance of Emancipation 119

CHAPTER 6

The Intrusive Nature of Mobile Telephony 123

Introduction 123
Mobile Telephony in Settings with Heavy Normative Expectations 125
Mobile Telephony in Interpersonal Situations 130
Initiation of the Call and the Production of Social Partitions 132
Management of the Local Situation During the Call 136
Reemergence into the Local Setting 138

Forced Eavesdropping and Being Embarrassed for Others 140
Conclusion 142

CHAPTER 7

Texting and the Growth of Asynchronous Discourse 145

Introduction 145
The Growth of Texting 149
Texting and the Individual 149
Texting and the Group 152
What Is Being Said, Who Is Saying It, and How They Say It 154
Content of the Messages 154
Mechanics of SMS Writing 157
Written vs. Spoken Language 162
Gendering of Text Messages 164
The Future of Texting 165

CHAPTER 8

Conclusion: The Significance of Osborne's Prognosis 169

Introduction 169
Interaction Between Innovation and Social Institutions 171
History of Technical Innovation and Social Adoption 172
Sociology and the Role of Technical Innovation 174
Social Capital vs. Individualism 177
Social Capital 177
The Institutionalization of Individualization 179
Role of ICTs in the Fostering of Social Capital/Individualism 181
The Internet 181
The Mobile Telephone 183
Ad Hoc Networks 187
Virtual Walled Communities 189

Appendix: Data Sources Used in the Analysis of Mobile Telephony 197

Endnotes 201

Bibliography 223

Index 239

Preface

The first time I can remember using a wireless communication system I was with "Doc" Scott, the local veterinarian. When I was in grade school, I had ambitions of becoming a vet, and since Doc Scott was a family friend, I occasionally got to ride with him on his rounds. He drove a white Ford Ranchero. The cab was littered with the paraphernalia of his trade, and in the back there was a box containing all his medicines and equipment. In addition, it had a big whip antenna and a two-way radio. The three local veterinarians used the two-way radio. With the help of their secretary/dispatcher, Mary, they coordinated their calls and arranged appointments as they drove from farm to farm in the countryside of north central Colorado. One day when I was riding with Doc Scott, he had a call out at a diary farm. It seems that he was awaiting some information from his office while at the same time he needed to attend to a sick Holstein cow. Since I was his assistant for the day, he asked me to sit in the car and wait for Mary's call. When she called, I was to go and get him. Thus, I sat there with a palpable sense of importance. I was doing something essential, and the execution of my assignment meant that I would have to use a new type of technology. After a few minutes, Mary called and said something like "I got that account. Over." In my debut as a mobile communicator, I fumbled through some answer and scuttled off to tell Doc Scott that Mary had called back.

The last time I used wireless communication was just now. I am sitting in my kitchen. It is a little past 5:00 in the morning; prime writing time. Since I cannot hook up to my e-mail without awakening the whole house, I have managed to configure my mobile phone in order to read my e-mail. This morning I see that I have spam from a woman named Henrettia who wants to sell me Viagra (50% discount!), an offer to buy some CDs from somebody called Melanie, a newsletter from the *New Scientist*, about a dozen e-mails having to do with different projects, and a note from a friend in the United States regarding a trip we are planning this summer—nothing I need to concern myself with until I get to work.

In between that early interaction with Mary via Doc Scott's radio and Henrettia's (soon to be rejected) offer lies a revolution in the development of mobile communication. The technology has moved from the rather limited use by special groups to being generally accessible.

In my youth, aside from that memorable experience with the veterinarian's radio, mobile communication was nothing if not fantasy. It was something one saw on TV "spy" programs such as *The Man from U.N.C.L.E.* and *Get Smart*. Like flying cars and X-ray glasses, these devices were nothing one would experience in the real world.

In the late 1980s, when I arrived in Norway, I experienced the next step in the popularization of the mobile telephone. I had been on a long flight from Denver via New York and Amsterdam to Oslo. The plane out of Denver was late, and as a result all the other connections had been squeezed. As you might expect in these situations, my baggage was lost. After going through the normal bureaucracy, I left the airport and waited at my friend's apartment. The next morning the phone rang. It was a delivery person from the airline, calling from the mobile telephone in his truck. He wondered about the specific address for the apartment so that he could deliver my suitcase. In addition to being relieved at getting back my toothbrush, the thing that fixes the event in my mind was that the truck driver was using a mobile telephone. It seemed extravagant beyond belief. I had heard of mobile phones, but I associated them with yuppies and other highfliers. Here was a normal fellow who was using one of the early "car-mounted" mobile phones for the simple task of finding an address.

My next brush with mobile telephony was during my job interview for my current employer. In 1993, I had applied for a job at the research institute for the Norwegian national telephone company. The institute's assistant leader attended the interview. Being on call, he had a mobile phone with him. I remember his coming into the room holding the device; a wonder of miniaturization. Indeed, it was so small that with some effort you could carry it in your pocket.

Though I did not know it at the time, this research institute had been a key partner in the development of mobile telephony as we know it today. People from the institute were involved in the development of the Nordic Mobile Telephone (NMT) standard, the first mobile telephone system that allowed for international roaming. They had also been active in the specification of the soon-to-be-commercialized Global System for Mobile communications (GSM) along with its Short Message System (SMS) function.

This was a lucky coincidence for me. As a sociologist, I was interested in how people dealt with technology. At that point, both mobile telephony and the Internet were on the verge of becoming popular. On the Internet side, the Mosaic browser had become available, and various news, billboard, and chat groups, along with MUDS and MOOS, were starting to flourish. At the same time, the mobile telephone was quietly reaching a broader and broader public.

There was a lot of sociological interest in the Internet. However, I came across several items that pricked my interest in mobile telephony and led me to study mobile telephony instead of the better-trod path of Internet analysis. The first of these was a simple chart showing the number of minutes per day spent on the telephone (all types) by Norwegians between the ages of 9 and 80. The line graph showed a dramatic rise through the early teens, with the high point coming in the mid 20s. From there, the line dropped slowly through the remaining age groups. The important thing was the near-vertical rise during the early teen years. My thought was that if any social indicator changes that fast in so short a period, there must be some form of social turbulence associated with the change. It indicated to me that adolescence was an important period when considering the adoption and use of telephony. You could guess that disagreements as to who could call and how long they could talk, as well as the unquenchable desire for teens to talk with each other, likely stood behind the statistics.

At about the same time I saw the transcripts from interviews done with mobile telephone users. This was before the adoption of the device by teens. There were, however, several young adults as well as some older users in the interview groups. Analysis of this data showed that the younger people wanted to use the mobile telephone to communicate while the older ones were more concerned with safety issues. There were also many comments as to the disturbing nature of the mobile telephone.

Based on these insights, we organized a series of in-home interviews in Norwegian homes in 1997 with teenaged families. It was our intention to understand the dynamics of the interaction surrounding the use of telephony and electronic communication. We interviewed the teens as well as their parents. As suggested by the charts, the interviews uncovered many strains and tensions. At this point, pagers were popular among teens. "Pay as you go" subscriptions — wherein you pay beforehand for calls and messages sent later — had not been commercialized, and mobile telephones were still relatively expensive. Pagers provided teens with a way to manage social networks that was beyond the control of their parents. This, in turn, resulted in discussions about the role of parents, their need to have control over their children, and problems associated with the cost of telephony.

The teens also noted that a few of their classmates had started to have mobile telephones. Not long after this, inexpensive terminals, along with "pay as you go" subscriptions, led to the widespread adoption of mobile telephones by teens in Norway and in other European countries. This was because the teens could easily afford the terminals and the subscription plan eliminated parental worries about excessively high telephone bills.

These elements told me that this was an interesting area. Aside from some work done by Lene Rakow and by Klaus Lange, there was little that had been done. Thus, it seemed like a natural area to examine. The scholars in the COST 248 action provided another strong influence. This group included Leopoldina Fortunati, Leslie Haddon, Roger Silverstone, Chantel de Gourney, Zbigniew Smoreda, and Enid Mante-Meier. This group showed me that there was broader interest in the issue, and it led to the development of other projects.

The work of the COST group also led to my meeting Jim Katz. He and Mark Aakhus, both of Rutgers University, organized a seminar and later published the proceedings in the book *Perpetual Contact*. In some ways, this was a watershed. It was confirmed that there is a lot of interesting sociological analysis to be had in examining with mobile telephony. The seminar, along subsequent seminars in Italy, South Korea, and now Norway, provided a forum in which we were able to examine how mobile telephony is used to coordinate activities and to provide a sense of safety and accessibility and how it disrupts the public sphere. In addition, it took the social analysis of mobile communications beyond Europe to the United States and Asia.

As we will see, the mobile telephone is more than simply a technical innovation or a social fad. The examination of its adoption and use and of the attitudes associated with the device provides insight into some of the broader machinations of society. In this process, the sociologist is provided with a rare opportunity to see the domestication of a new technology and its various consequences.

This book considers the various dimensions of the phenomenon. Cultural differences, differential access to equipment, alternative-pricing systems, and different needs are all elements that can play out in various ways. Within Europe there are definite north/south, age, and gender differences in ownership and use. Going somewhat further afield, I-mode, in Japan, and the experience of Korea, the Philippines; and Taiwan point in one direction; the experience of mobile telephony in the United States points in another. Finally, the perhaps unique experience of, for example, Grameen Phone in Bangladesh provides a third insight into the role of mobile telephony and the range of trajectories that are possible.

Much has changed between the time of my conversation with Mary over Doc Scott's radio and the ability to read my e-mail via a mobile telephone. There has been the development and adoption of a major new form of communication around the globe. Although the technology is in place, we are still unsure as to the social consequences of mobile communication. This book is an attempt to consider its impact on our everyday lives. But first, I have to erase that Viagra spam ...

Acknowledgments

There are several people and organizations I would like to thank for their contribution to this book. First is my patron, Telenor. In particular, Telenor R&D and Telenor Mobil have been generous. They have supported my work and have been munificent in their support of various data-collection projects. My colleagues Birgitte Yttri, John Willly Bakke, Per Helmerson, Kristin Braa, Tom Julsrud, Siri Nilsen, and Kristin Thrane have all contributed to this work in various ways. Birgitte deserves special recognition for her insight and her willingness to toss about ideas and to think through sociological issues. Zahra Moini also deserves special mention as one who was wise enough to focus on the social consequences of mobile telephony.

Naomi Baron of American University has, more than any other, been the midwife of this book. She has been an irreplaceable support and advisor. Her comments, generous reviews, and willingness to provide assistance and her kind words are warmly appreciated. If Naomi was the midwife, Leysia Palen of the University of Colorado nurtured the book through its development. Leysia's judgment, insight and commitment have helped to refine and focus the work. Her tireless engagement in the reviewing of the manuscript has been a godsend.

My colleagues Leslie Haddon, Leopoldina Fortunati, Christian Licoppe, Kjell Olav Mathisen, and Bella Ellwood-Clayton, provided excellent reviews, helped me to correct the shortcomings of the various sections, and provided me with insights that have enriched the text. I thank them for this.

I have had the invaluable benefit of extensive comments provided by Scott Weiss of Usable Products and Martha Lindeman of Agile Interactions. They have been thorough and insightful in their reading of the various drafts. Their comments have been some of the most useful I received.

A special thanks to Troy Lilly and Elliot Simon or their diligence in the preparation of the manuscript.

I also want to name my colleagues in Norway, Europe, and the United States. In particular, the activities within the COST 248 project, the EURESCOM P903 project, and the EU Youngster project have been positive and helpful in the development of the ideas described here. In this context Enid Mante-Meier and Leila Klamer as well as Annvi Kant should receive special mention. I am indebted to Jim Katz for his support, his comments, and his never-failing advice. I would like to thank Jonithan Grudin of Microsoft Research and Diane Cierra for their help in the development of this book.

Finally, I cannot adequately thank my wife Marit and my daughters Nora and Emma for their support and their willingness to let me occupy a corner of the kitchen with my papers, PC, and other assorted paraphernalia while working on this tome.

1

CHAPTER

Introduction

Introduction

Not long ago I was looking through some of my grandfather's papers when I came across a newspaper clipping from the front page of the *Denver Post* dated August 20, 1946. The main upper fold headline, in all caps, was "RADIO-PHONE HOOKUP BEGUN IN COLORADO." The article described an experimental project carried out by Mountain Bell, then the local telephone company in Colorado. It dealt with the provision of radio-based telephony to residents of Cheyenne Wells, a small town on the high plains of eastern Colorado, near the Kansas border (Figure 1.1). The article started by stating: "For the first time anywhere in the worldwide Bell system, that grand old institution, the party line 'went radio' at Cheyenne Wells, Colo., Tuesday noon." While there had been radio-based communication using Marconi's wireless radio for some time prior to this, here was radio-based telephony being applied to the needs of common individuals. The technology was not being used to send messages from ships or to broadcast baseball games, as in the case of commercialized radio stations. Rather it was being applied to the mediation of interaction between private individuals. Indeed this was one of the ancestors of modern radio-based mobile telephony.[1]

My grandfather was an engineer in Mountain Bell and an early radio enthusiast. While at college at the University of Colorado in 1919, he built what was probably one of the first radio devices in Boulder, stringing an antenna between the old engineering building and the smokestack of the power station. He participated in the Cheyenne Wells project as the chief radio engineer; and thus the archiving of the article among his papers.[2]

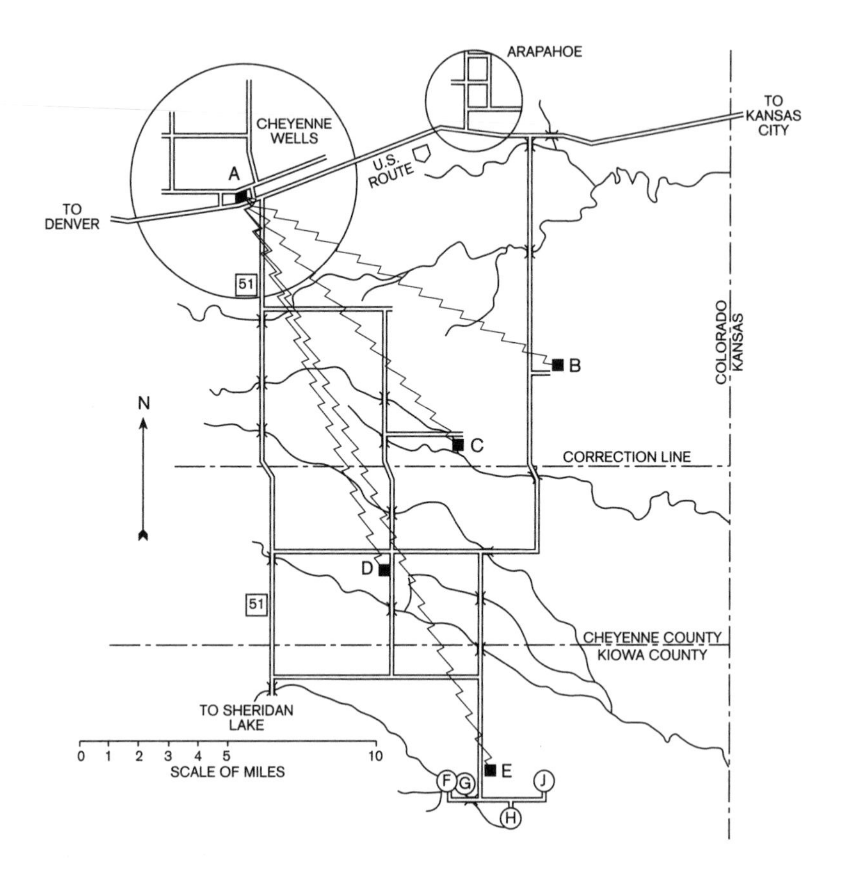

FIGURE 1.1 Map of the region served by the radio telephone system in Cheyenne Wells, Colorado, August 1946. Source: Originally published in Moore, Seyler, and Wright, 1947.

If I know my grandfather, he was far more excited by the details of the technology than by its impact on the lives of the farmers living in Cheyenne Wells. His papers from that period are full of technical drawings of circuits, switch interfaces, etc. Radio technology was, after all, a way to quickly establish a telephone service without having to go through the laborious process of building the physical telephone lines.

Nonetheless, local access to telephony was nothing if not revolutionary for these people living on the high plains. Claude Fischer provides an account of how early telephony had perhaps a more profound impact on rural communities than on more urban locations. He tells how the landline telephone — sometimes even using barbed wire fences in lieu of traditional wiring — were used when "calling for help in emergencies, obtaining weather forecasts and crop prices, ordering goods, recruiting temporary labor, and so on" (Fischer 1992, p. 98).

If we move our locus from the plains of eastern Colorado to the Chittagong region of Bangladesh and shift in time from 1946 to 2003, we find a similar type of development under way. Using radio senders along the local train tracks, Grameen telephone in Bangladesh has used its system of no-collateral microloans in order to distribute mobile telephones with solar-based battery rechargers to women living in rural villages. While not ignoring the fact that Bangladesh has one of the lowest adoption rates in the world — less than one telephone subscription per 100 persons — like the project in Colorado, access to telephony has the potential to revolutionize the lives of the local villagers. It provides them with direct access to agricultural commodity pricing and thus provides improved bargaining power vis-à-vis commodity wholesalers. In addition, it allows the management of payment and transfer of funds, access to medical services, contact with distributed family members in the case of family emergencies and medical situations and on various social occasions, aids in the arrangement of diverse official papers, etc. (TDG 2002; Singhal 2002).

If we move the scene once again, this time to my location in Oslo, Norway, as I write these words, we find other consequences of mobile telephony. As opposed to the relatively functional use of the device as seen in the previous situations, we find a thriving culture of mobile telephony in which the mobile telephone is used to coordinate peoples' everyday comings and goings. It is being used to chat. It is being used by teens to send text — and even picture — messages. It is being used to call sick aunts and by parents to organize children's birthday parties. It is being used by lovers to exchange endearments. It is being used by businesspeople to buy and sell. It is being used by tradespeople to keep track of their assignments and to assist them in their work. It is being used to surf the Internet and to send multimedia messages. It is being used to give people last-minute directions and to make last-minute arrangements. It is being used to delay or rearrange meetings. And, sometimes simultaneously, it is bothering others, who happen to be within earshot, almost to tears.

From Japan to the Philippines, Europe, North America, and beyond, the mobile phone has announced its presence. We see — and hear — it in unexpected locations at unexpected times for unexpected reasons. In Scandinavia, Italy, Israel, Korea, and Japan, it is common to see people chatting on their mobile phones as they walk down the street. In the United States, people are using up their nationwide-whenever-wherever-anytime minutes to keep in touch across time zones. Teens — who are the archetypal mobile superusers — "text" to each other quite literally throughout the day and night. Plumbers, carpenters, and other blue-collar workers whose place of work shifts from day to day have found that mobile telephones allow them to work more efficiently and to blend their work and private lives. Business- and tradespeople use the device to make their days more effective. People use it, perhaps unwisely, while driving their cars, and others make calls while on public transport — to the eternal annoyance of their fellow passengers. In short, it is being used to spin the web we call the social network. The scene in the street immediately below me is also being played out in other "mobile" cities and countries. Teens in Rome, Manila, and Seoul, mothers in Jerusalem and Den Haag, and businesspeople in Tokyo and Helsinki are all using the mobile phone in the course of their everyday lives.

In the early 1950s Harold S. Osborne — the recently retired Chief Engineer for AT&T — predicted that mobile telephony would eventually allow us ubiquitous access via small portable devices. He foresaw that anytime we wished to talk to another, we would simply use a small device to punch the appropriate number. He thought that these devices would allow us to hear the voice of our friends and to see them in three dimensions (Conly 1954).

We are beginning to see the fulfillment of his prognosis in a technical sense. But obviously, all of this has not happened without controversy. The mobile telephone's invasion of restaurants, buses, parks, and even public toilets has given new meaning to the concept of eavesdropping. Smith noted in a *Readers Digest* article from 1937 that "there is no room in the house so private that he cannot crash it by telephone" (Smith cited in Fischer 1992, p. 225). The mobile telephone extends this lament to the far reaches of civilization, and, as many suggest, beyond.

Although people complain of the mobile phone's intrusion into polite society, there are many other aspects to this phenomenon. We rely on the mobile telephone. It helps us coordinate our lives while on the run; it provides us with a sense of safety and gives us accessibility to others. We personalize the device, and in doing so we make a statement as to who we are and how we want to be seen. It is also worth noting that this transition has come quite quickly. A decade ago the mobile telephone was the symbol of yuppies, not teens. Now it has nudged and pushed its way into our everyday lives in new ways. The rise of mobile, push to talk, multimedia messages and various handheld computing devices will add a new twist to all this.

This book looks into the social consequences of mobile telephony. The particular focus is on the private sphere.[3] It examines how the mobile phone is used to provide a sense of safety, to coordinate activities, and to provide accessibility, and it studies how this device disrupts the public sphere. The mobile phone's appearance in society has resulted in turbulence. In this process it has exposed the "taken for granted" assumptions regarding how society and social institutions function. It provides insight into how adolescents manage or — to use the concept developed by Glaser and Strauss (1967) — how they shape their emancipation. It is used in the integration of the teen peer group, it allows for "real-time" microcoordination of social interaction; it "softens" appointments and questions our assumptions about the sanctity of formalized meeting times; it engenders new notions of safety and security; it enables us to colonize portions of the public sphere for personal interaction; and it even exposes the elasticity of language, as seen in the rise of Short Message Service (SMS) and texting. At a slightly more abstract level, we can see that the mobile telephone lowers the threshold for social interaction within groups and, at least within the groups, contributes to the maintenance of social capital.

Thus, the mobile phone is more than simply a technical innovation or a social fad. The examination of its adoption and use, and of the attitudes associated with the device, provides insight into some of the broader machinations of society. In this process, the sociologist is provided with a rare opportunity to see the social adoption of a new technology and its various consequences (Silverstone and Hirsch 1992; Silverstone and Haddon 1996; Palen *et al.* 2001).

My window on this development has been that of an expatriate American in Norway, where I live and work. Along with the other Nordic countries, Norway has been quick to adopt and use the technology. While there are special issues and adaptations associated with the Nordic scene, the quickness with which mobile telephony was adopted and the ongoing love/hate relationship with the mobile phone perhaps provide insight into the broader social dynamics of the device. Thus, in Norway I can observe a relatively mature mobile culture. Norway was one of the countries supporting the development of the Nordic Mobile Telephone (NMT) standard that allowed for international roaming. In addition, it participated in the specification of the Global System for Mobile Communications (GSM) standard. It is also a country that has taken mobile telephony into everyday life and is indeed among the countries with the highest rates of ownership. Thus, it gives insight into the benefits and problems of mobile communication.

Beyond the Scandinavian experience of mobile telephony, this book considers the cross-cultural dimensions of the phenomena. Cultural differences, differential access to equipment, alternative-pricing systems, and different needs are all elements that can play out in various ways. Within Europe there are definite east/west, age, and gender-based differences in ownership and use (Mante-Meijer *et al.* 2001; Ling *et al.* 2002). Beyond Europe, the experience of I-mode in Japan, the rise of texting in the Philippines, and the intense use of the mobile telephone in Korea provide us with an alternative perspective. The somewhat more cautious use in the United States acts as a type of counterpoint. A completely different picture regarding the diffusion of mobile telephony arises when we look at the experience of, for example, Grameen Phone in Bangladesh.

This book will look into these social consequences of mobile telephony. It will consider the impact of the device on our everyday lives. It will examine how we are domesticating the device and how it is changing the way we consider issues such as accessibility, coordination, safety, and teen use of mobile telephony. It will consider why texting and SMS has grown to be a dominant service and study the role of the mobile telephone in the public sphere.

History of Mobile Telephony

Contemporary mobile telephony is an offshoot of the more general development of radio communication that started in the late 1800s. Following on the work of Maxwell, Hertz, Hughes, and others, Marconi became central in the development of radio-based communications (Farley 2003). Through the 1890s, he sent radio signals over progressively longer distances, ranging from a several hundred meters to several kilometers and eventually to transoceanic communications. By 1899, Marconi was able to equip two ships with radio transmitters in order to report the progress of the America's Cup. Two years later, he successfully sent a radio message from Cornwall, England, to Newfoundland, Canada.

This new form of communication grew and developed during the first years of the 20th century. The growth of radio communication was aided by De Forest's development in 1906 of a vacuum tube, which allowed for the amplification of radio signals.

Marine communications was one of the first areas of truly mobile radio-based communication (Grimstveit and Myhre 1995; Haddon 1997). During this period, passenger ships, fishing fleets, and freighters were regularly outfitted with radio equipment. The Titanic disaster in 1912 led to the requirement that passenger ships maintain 24-hour radio watches.

Radio telephony, that is, the integration of radio transmission with the traditional switched telephone network, was somewhat superficially examined in this period (Brooks 1976). However, the new medium was generally channeled into other areas. In addition to marine communications, the period saw the development of commercial broadcast radio. In Detroit in the early 1920s, radio communication was used to coordinate the activities of police, taxis, etc. (Manning 1996; Farley 2003; Dobsen 2003).

The development of the transistor after the Second World War led to the next significant development in mobile telephony, namely, the regular use of radio telephony for switched communications. From the late 1800s, landline telephony had offered person-to-person communication via switched circuits. A dedicated circuit, or a "line," was set up between two persons calling each other. The early "switches" were simply manual devices where the operator (usually a woman) determined who the calling party wished to reach and then connected the two with the use of a cable and a jack. These switches were progressively automated until switching systems are now largely electronic devices that can handle many thousands or even millions of simultaneous conversations.

By contrast, radio-based communications, such as those carried out by police, firefighters, taxi companies, and the like, often involved — and indeed in many cases still involve — the use of a central dispatcher who directed communications to the appropriate person and who also maintained an overview over the activities of all the actors in the group (Manning 1996). The messages are broadcast, and there is no dedicated circuit available for personal communication (Figure 1.2).

As evidenced by the work in which my grandfather participated, the wide-scale integration of radio-based telephone devices with traditional switched telephony systems started in the late 1940s. AT&T extended this to mobile telephone systems. By today's standards, these mobile devices seem like Rube Goldberg machines. The person placing the mobile call had to manually search for an unused channel on the radio telephone. The individual then used that channel to contact an operator, who in turn actually dialed the number provided by the caller. When the person called came on the line, the connection was half duplex, meaning that only one person could speak at a time (Forley 2003). A "push to speak" switch on the handset opened and closed the channel for the caller. This controlled the pace of the conversation. In order to hear the response of the person called, you would have to release the switch. Presumably, the conversation would include conventions such as "copy," "over," and "roger," in order to facilitate taking turns.[4] Interestingly, the development of "push-to-talk" systems are reviving this one-way form of interaction.

FIGURE 1.2 An early example of mobile, voice-based radio communication. Copyright © 2000 Lucent Technologies. http://www.bell-labs.com/history/75/gallery.html

By the mid-1960s an improved system was developed. In this case, there were automatic channel assignment, direct dialing, and full duplex operation. The system in a specific geographic area allowed for only about a dozen simultaneous users. In 1976, the system in New York City, for example, had almost 550 users sharing 12 lines. There were 3700 customers on a waiting list (Encyclopedia Britannica 2002). The mobile terminals themselves were ponderous things requiring batteries heavier than a car battery. Thus, mobile telephony at this point often meant automobile-based telephony.

The 1980s saw increasing interest in the development of various mobile telephony standards. In the United States, these included the mutually incompatible Advanced Mobile Phone System (AMPS), Narrowband Advanced Mobile Phone System (NAMPS), Time-Division Multiple Access (TDMA), and the Code-Division Multiple Access (CDMA). These systems progressively allowed increasing capacity as mobile telephony became more popular.

In Europe, the Nordic Mobile Telephone (NMT) was the first generally successful cellular system that automated the calling process and allowed for international roaming. The system was established in the early 1980s in Sweden, Denmark, Norway, and Finland. Since it was a standardized system, one could use the same mobile telephone across the whole region. Nonetheless it was still a parochial system. It was incompatible with the Total Access Communication System (TACS) in the United Kingdom, the Radio Telephone Mobile System in Italy, RadioCom in France, and a number of systems used in other countries. Because of this incompatibility, in the late 1980s the European public telephone network operators, in conjunction with the European Community and the European Telecommunications Standards Institute (ETSI), started the development of the GSM. This digitally based standard has come to dominate the world's mobile telephone market. GSM allows for international roaming, is backward compatible with other systems, allows for various national tariff systems, and includes the ability to send and receive various data-based services, such as the much-maligned Wireless Application Protocol (WAP) and the much-adulated Short Message System (SMS). In addition, it includes items such as caller ID, call waiting, and voice mail.

GSM was an immediate success in many European countries. In the early and mid-1990s there was nearly a "Klondike"-like feeling around the marketing of GSM, with adoption rates continually exceeding the budgeted expectations. For example, TeleDanmark hoped for 15,000 new customers in 1993 but got more than 65,000. Sonofon planned its GSM system for 25,000 customers in 1995 but achieved 100,000 (Haddon 1997). While original expectations were often framed around business markets, mobile telephony has very quickly moved into the private sphere. This development was aided by the marketing of subsidized handsets, where one can purchase a handset and a subscription for a very low price. The development of prepaid or "pay-as-you-go" subscriptions also encouraged the adoption of mobile telephony. Indeed, its widespread adoption by certain groups, such as teens, has been greeted as a mixed blessing in many quarters. Thus, according to the International Telecommunications Union (ITU), as of 2003 almost 69% of all mobile telephone subscribers in the world were using the GSM system. The various others standards make up less than one-third of all subscribers.

At about the same time that the GSM standard was being commercialized, handset manufactures were radically reducing the size and weight of mobile telephone terminals. Rather than being the large "lunchbox" affairs that were best mounted in your car, they begin to be devices that could conveniently be placed in your pocket. Functionality developed to include things such as WAP, SMS, call logs, overview of appointments, and contacts. The terminals included color screens, gaming functions, and the capabilities to download ringing sounds and logos and to shift the covers. Battery capacity increased and still other functions have appeared, such as cordless microphones/earphones and integrated cameras.

WAP emerged in 1997. It was an effort to allow Internet-like services within the GSM system and to avoid a situation where different commercial actors would develop separate standards. This was based on the cooperation of Phone.com, Nokia, Ericsson, and Motorola to produce a license-free protocol. WAP was commercialized in 1999 amid a great deal of hype. It has not managed to live up to hopes because of its limited usefulness, the limited performance of the networks, and the lack of terminals at the point of commercialization. WAP also suffered in comparison to the proprietary I-mode system. With the introduction of higher-speed General Packet Radio Service (GPRS), broader access to WAP-capable terminals, and more sober expectations, WAP has found a somewhat limited role in the mobile telephone firmament (Lindmark 2002).

The final system of note here is the DoCoMo I-mode system in Japan. I-mode provides access to a variety of services and allows one to send and receive short messages as well as e-mail.

Another unique element in the I-mode system is that it allows access to a broad variety of premium information services. The I-mode standard specifies a "lite" version of HTML, the coding used on the Internet. This facilitates the development of Web-like sites. The other unique aspect of I-mode is that the parent company, DoCoMo, oversees and administers the available sites. Thus, we cannot really speak of an open Internet, but rather of a type of extended intranet where one has access to a large but cosseted set of services. DoCoMo provides developers with a type of style book that guarantees a similar layout for the different sites.

The user has access to a variety of both free and subscription services in addition to the traditional voice and text communication functions of other mobile telephony systems. DoCoMo handles the billing and charges a 9% fee for all revenue generated by the various sites. In addition, DoCoMo receives the standard payment for transmission of data through its network. Thus, I-mode is not so much a technology as a system of infrastructure and marketing.

Mobile communication is moving in the direction of broader types of access and new forms of communication. Wireless local area networks, handsets that include Internet browsers, and increased speed in the network are being commercialized. In addition, new forms of messaging that include the exchange of photographs and sound are moving into the market. In spite of this, the fundamental services of person-to-person communication, based primarily on simultaneous voice communication but also on asynchronous text messages, are central to the popular use of mobile telephony.

Finally, mobile telephony allows for the development of location transmission services that, in turn, open the way for geographical positioning, the ability to find the location of various services, etc. Imagine being able to call up a map showing the physical location of your friends (hopefully only those friends who have given permission) (Ling 2002).

Thus, mobile telephony has grown from being a rather ponderous and awkward system to being an easily transported part of everyday life. Its functionality has grown beyond simple communication to a system that allows for the communication of text, access to the Internet, the capturing and sending of images, and the distribution of location-sensitive information.

Growth of the Mobile Market

Along with technical developments, the number of subscribers to various mobile telephone services has seen a dramatic growth since the early 1980s. As of 2003, there were approximately 1.162 million mobile telephone subscriptions (ITU 2003). On a worldwide basis, there is, roughly speaking, one mobile telephone subscription for every fifth or sixth person. To put this into some perspective, there are slightly more mobile telephone subscriptions than traditional landline subscriptions. In terms of distribution (see Figure 1.3), roughly one-third of all mobile telephones are in Europe, almost 40% are in Asia, and slightly less than one-quarter are in the Americas. About 3% of all mobile telephones are found in Africa and 1% in Oceania.

In order to place mobile telephony into a global context, we can look at the adoption rates for various regions and countries. Before embarking on this analysis, however, there are several caveats. First, the per capita number of subscriptions is simply the total number of subscriptions compared to the total population. It ignores the fact that in some cases subscriptions are associated with functions as opposed to individuals (e.g., ambulances or other mobile emergency vehicles). Further, per capita adoption rates ignore the fact that some individuals have more than one subscription. For example, in 2000, about 13% of all teens in Norway had two or more mobile telephony subscriptions.[5] Finally, there are often many "dead" subscriptions and discarded handsets. This is particularly the case with prepaid or "pay-as-you go"

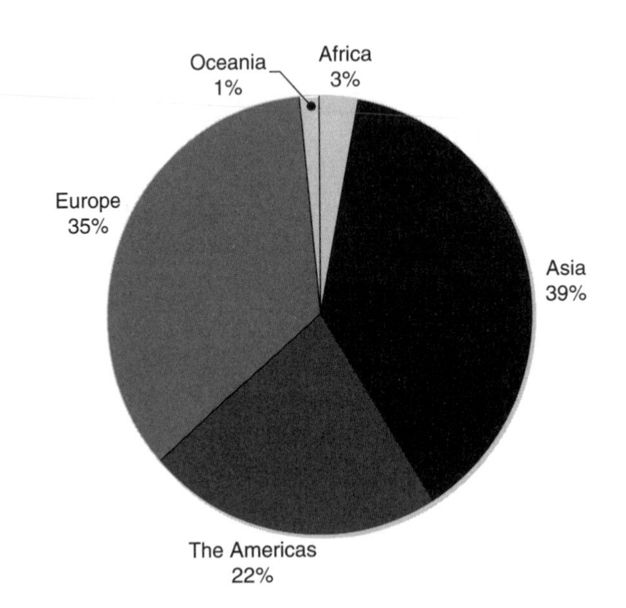

FIGURE 1.3 Worldwide distribution of mobile telephones, 2002. Source: ITU.

subscriptions, where the subscriber has paid for use beforehand. If the subscriber does not actively use the telephone, that is, if no calls are placed for six months or a year, the network operator wishes to get the subscription off its books. Thus, the subscription eventually becomes invalid. The point at which these are considered inactive is not clear, which often slightly inflates the statistics in those countries where there is an extensive number of prepaid subscriptions.

Thus, while the ITU reported 82 mobile telephones per 100 persons in Norway in 2001, a survey I carried out during the same period showed that 75% of those asked said they had a mobile telephone, while another 8% said that they could borrow one on a more or less regular basis. If we assume that two persons shared the "common" mobile phones, that comes up to coverage of only about 79%. Thus, the remaining 3% reported by the ITU comprise either the "function"-related telephones, those who had several subscriptions, or the dead and dying subscriptions that were counted as living.

Given these caveats, the ITU material shows that there were about 18.8 mobile telephones per 100 persons in the world as of 2002. About 22 countries, largely in Europe, have adoption rates of over 75% (Figure 1.4). The material also shows that there are approximately 29 countries with less than one mobile telephone per 100 persons. These were clustered in Africa and in parts of Asia.

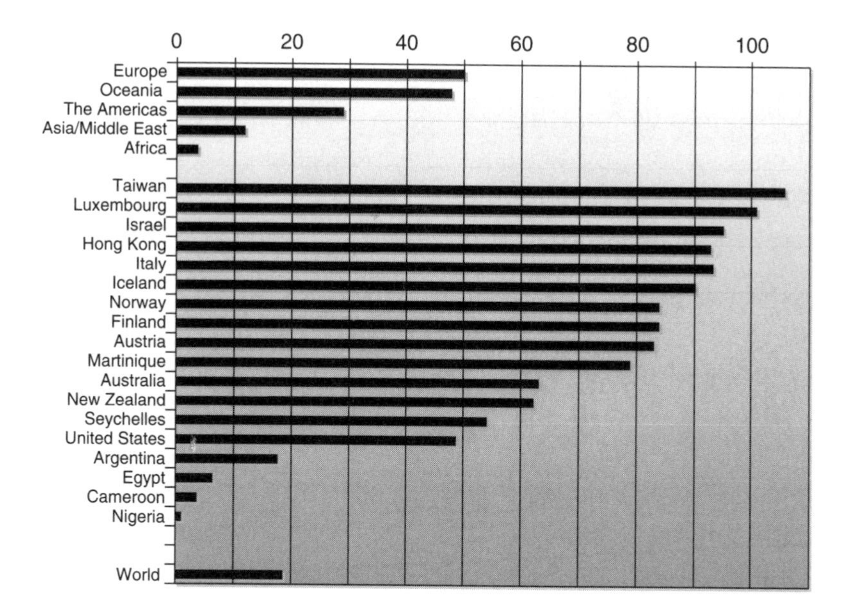

FIGURE 1.4 Mobile phones per 100 persons for various regions and countries, 2002. Source: ITU.

Europe is the region with the highest adoption rate in the world. The material from the ITU shows that in Europe there were slightly less than 50 subscriptions per 100 persons. There are 14 countries with a subscription rate of 90% or higher. These were Iceland (90.28), Italy (92.65), and Luxembourg (101.34). There were also 13 other countries with an adoption rate of more than 75%. Interestingly, the data shows that Finland (84.5), which has the image of being the most advanced mobile country in the world, was slightly behind the top tier of countries. While there are many countries with high adoption rates in Europe, these were counterbalanced by many Central and Eastern Europe countries, such as Bosnia (9.17), Russia (12.05), and Ukraine (4.42), where a much lower adoption rate is found.

The region with the second highest adoption rate — albeit spread across a relatively small population — was Oceania (48.53). In this region, Australia (63.97) and New Zealand (61.84) are clearly the most advanced mobile countries and are the major population centers.

In the Americas, the material from the ITU showed that there were about 30 mobile phone subscriptions per 100 persons. Interestingly, Martinique (78.99) and Guadeloupe (69.72) headed the list of the most mobilized countries in the region. The United States was third, with 48.81% coverage.

There were 12.19 mobile telephone subscriptions per 100 persons in Asia/Middle East. The heavyweights in this region were Hong Kong (92.98), Israel (95.45), and the country with the highest adoption rate in world, Taiwan (106.45). Taiwan edged out Luxembourg by about 5 percentage points here, but when one considers that there are more than 22 million Taiwanese, compared to only 450,000 Luxembourgers, you can perhaps appreciate the magnitude of mobile telephony in the former. It is worth noting that the rates for Japan (62.11) and South Korea (67.95) place them somewhat back in the pack.

As with Europe, there are also many countries with relatively low rates of adoption. China (16.09), India (1.22), Indonesia (5.52), and Bangladesh (0.81) are examples of large population groups where there are few mobile telephone subscriptions. The sheer size of China means that it is the single largest mobile market in the world, with about 200 million subscriptions, 130 million of which are GSM. Nonetheless, there are only about 16 subscriptions per 100 persons there.

For Africa as a whole, the ITU material shows that there was 4.19% coverage in 2002. The thing of note here is that there were often nearly as many mobile as landline subscriptions. This statistic describes both the relatively quick adoption of mobile telephony and the impoverishment of the African landline telephony system. In Cameroon, for example, there were slightly more than 4.4 landline telephones per 100 persons. When it comes to mobile telephony, Cameroon had just over 3.5 mobile telephone subscriptions per 100. By way of comparison, in the United Kingdom there were something like 144 landline telephones per 100 persons vs. 84.49 mobile telephones. The landline-to-mobile ratio in the United States was 114 landline phones to 48.81 mobile phones, and in Luxembourg, which along with Taiwan seems to be some sort of telephonic Mecca, there were a whopping 178 landline telephones vs. 101.34 mobile phones per 100 persons. Thus, while there are literally orders of magnitude between the situation in, for example, Cameroon and than in Luxembourg or Taiwan, it seems that mobile telephony is making inroads into those countries where there has been little telephony before. In addition, it is allowing for alternative types of access where one was telephonically isolated before, as seen in the case of Bangladesh.

In countries with high adoption rates, an increasing number of persons have only a mobile subscription. In 2001 about 8–10% of those persons in Italy, Norway, and the United Kingdom reported that they had only a mobile telephone. These persons were generally younger than those who had both a landline phone and a mobile telephone. Those who had only a landline telephone were significantly older than the other two groups.

There are several other elements that have favored the adoption of mobile telephony in countries and perhaps hindered it in others. These include pricing, system interoperability, and the coverage of mobile telephony (Robbins and Turner 2002). In terms of pricing, the system of "calling party pays" is not observed in the United States. Instead, the cost of a call is shared between the caller and the party called. This means that a cost is imposed on the person you call, as opposed to the system observed in other parts of the world, where the caller assumes the cost of calling.[6] Thus, in addition to taking up your time, receiving a call has economic consequences for the party called. From the perspective of the caller, this system might be seen as a psychological deterrence to mobile telephone use, since a caller might hesitate to impose himself or herself on another person both temporally and economically.[7]

In addition, system incompatibility can limit the usefulness of mobile communication. This means that if you, for example, have a GSM telephone in an area where there is only CDMA coverage, you will be unable to make calls. Finally, there is the issue of coverage. There are, for example, vast stretches of territory in the United States that are outside the range of coverage from any system.[8] In this respect, the denser population coverage of Europe and parts of Asia facilitate the development of mobile communications systems.

Beyond the raw adoption per capita, we can see that the mobile telephone is adopted unevenly across various groups in a particular country. If we compare ownership in the United States and in Norway (Figure 1.5), we see that as of 2002 the penetration rate was higher in Norway than in the United States. There were also several other interesting differences. In Norway, young adults had higher adoption rates than other age groups. By contrast, in the United States, the highest adoption rates are among mature adult groups. Another difference is that in Norway, mature men generally have greater access than the same-aged women. Many of these differences are likely due to employers subsidizing mobile telephone use. The same does not seem to be true in the United States, where both genders have roughly the same rates of access.

When considering social class and the adoption and use of mobile telephony, it seems as though there are not the same types of "digital divide" issues that we find associated with the PC and the Internet.

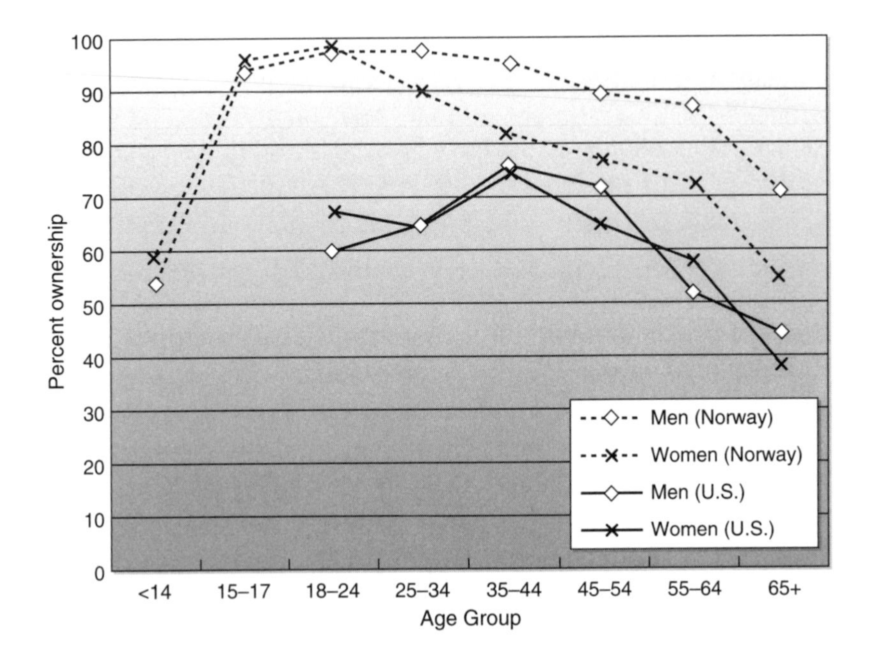

FIGURE 1.5 Reported mobile telephone ownership in the Norway and the US, 2002. Sources: Telenor, Pew Internet and American Life Project, Oct 2002 Tracking Suvey.

This said, there are clearly international differences in adoption rates, as we have seen. Further, within the context of various countries, analysis shows that there are some educational and income-based differences in access to and use of the mobile telephone. In a European-wide analysis carried out by Eurescom in 2000, the results show that there were significant income-based differences in ownership of the mobile telephone.[9] There were also differences based on educational attainment. In their analysis of the scene in the United States, Katz and Rice find somewhat similar findings (2002, pp. 253–254).

Nonetheless, mobile telephony is, in itself, more accessible than the PC/Internet. One does not need to deal with assembling various components and software. At the level of the user interface, the mobile telephone operates in much the same way as the traditional telephone: that is, you dial a number and the call is put through. While there are more advanced functions available — such as multimedia messages and Internet chat — you can use the mobile telephone without having to learn about or deal with these functions. In addition, the technology is relatively inexpensive and widely available. In some countries, you can buy a mobile telephone handset that includes a prepaid subscription at a gas station or in the grocery store for less than $40. Thus, you avoid the need to purchase expensive components that may or may not function flawlessly together.

Outline of the Book

In the remaining chapters, I will cover several areas, including (1) the role of the mobile telephone in fostering a sense of safety and security, (2) its use in the microcoordination of everyday life, (3) the role of the mobile telephone in the lives of teens, (4) the mobile telephone as a disturbing influence, and (5) the rise of SMS as a form of communication.

Chapter 2 examines the broader questions associated with the adoption and use of mobile communication. It examines the ideas of technical determinism (the notion that technology forms society), social determinism (the opposite idea: that society forms technology), and the two middle positions of the affordances approach and the domestication approach. In addition, the chapter discusses the methods and data that lay behind the analyses presented here. This discussion will be of interest to the theory and method wonks among us.

Chapters 3 through 5 follow a slightly Maslowian development, in that the discussion moves from a discussion of safety and security (Chapter 3) via a discussion of the functional use of the mobile telophone for coordination (Chapter 4) to a discussion of identity building and something approaching self-realization among teens (Chapter 5). A chapter follows (Chapter 6) that describes how the mobile telephone is a disturbance in the public sphere and that analyzes texting.

Safety and security (Chapter 3) is a common theme in the purchase and ownership of a mobile telephone. Indeed this is one of the basic functions of the device. Studies in Norway, Europe, and other countries all point to this conclusion (Mante-Meijer *et al.* 2001). This is often couched in terms of emergencies, but we find a range of situations here. The mobile phone, for example, gained a certain prominence in the tragic events of September 11, 2001. At a more limited level, middle-aged users spoke about being able to provide help in the case of car accidents, retirees with potentially dramatic health conditions talked about feeling free to use relatively remote mountain cabins because they can call for assistance in case of problems, and teens talked about being able to call home when they missed the last return bus. While the device allows us to summon help in certain situations, it can also be the cause of insecure situations, as seen in the studies of phoning while driving. Although the issues are quite broad, analysis shows that safety and security are possibly the best-recognized aspects of the mobile telephone.

The ability to coordinate activities "on the fly" is, perhaps, one of the most central advantages of the mobile telephone. It has led to new ways of organizing everyday life. It provides flexibility in the way we plan our days and in the way we use the transportation system. In its broadest sense, it can well be that the mobile telephone will lead to changes in the organization of urban life. The way one plans travel, the way meeting times are arranged, and the way we coordinate the various daily activities within the family at work are all part of this issue. The concept of "microcoordination" can be used to describe the nuanced use of the mobile phone for planning activities. I will examine this in Chapter 4.

Beyond the functional use of the mobile telephone, the device has led to a different understanding of interaction and networking among teens (Chapter 5). Teens in Scandinavia, Italy, Japan, Korea, and many other countries have adopted the mobile telephone, often to facilitate their social interaction. While many of the interests and activities are the same, the way in which they are organized is different. The device allows for a type of anytime-anywhere-for-whatever-reason type of access to other members of the peer group. This means that the social network is more tightly bound together and dynamic in its organization and location. Thus, it is clear that the mobile telephone has redefined the institution of adolescence as well as the emancipation process.

The discussion surrounding the disturbing use of the mobile telephone in public places is, conceivably, the classic gloss of the device in the popular imagination. This will be discussed in Chapter 6. The ringing of a mobile phone in a bus or a restaurant is the butt of many bitter comments. Indeed, the ringing of a mobile phone at a funeral has become a type of urban legend. It is a theme that almost without fail arises in focus groups, interviews, and even casual conversation. What is it about the public use of the mobile telephone that grates on the common imagination? What does this say about the technology, about the people who use it in public spaces, and about the role of public space in our social lives?

Upon the ringing of a mobile telephone, you must, in one way or another, excuse yourself from the face-to-face world and give yourself over to a telephonic sociability. This leaves the "precall" social situation "on hold" while you complete the call, which itself can require greater or lesser levels of privacy. Upon completion of the call, you must reintegrate yourself into the preexisting face-to-face social interaction and perhaps repair the damage caused by the interjection of the call.

Texting is the focus of Chapter 7. Since the late 1990s the use of short messaging or texting, has seen phenomenal growth. It is estimated that approximately 280,000 such messages are sent every hour in Norway, a country with only 4 million inhabitants. Among teens, this is the preferred form of interaction. Texting has several characteristics that make it useful not only for teens but increasingly for other groups. It is asynchronous (that is, it does not require the immediate attention of the receiver), it is relatively unobtrusive, and it is generally cheaper than voice telephony. Obviously, it suffers from the facts that it is somewhat difficult to enter the messages and that message length is limited to 160 characters. Nonetheless, experience shows that this has not been a major hindrance. Beyond texting, mobile telephones are also turning into multimedia terminals that allow us to send and receive music, pictures, drawings, and the like. These issues will be considered here.

Finally, in Chapter 8 the various threads associated with the use of mobile telephony in society are tied together. This chapter considers the current situation of mobile telephony and discusses prognoses about continued development, both in terms of technology and in terms of the realms in which the device will be used. The chapter summarizes the degree to which the device has become domesticated and, indeed, the degree to which this approach is applicable. In addition, it summarizes the broader social consequences of the mobile telephone, that is, security, coordination, accessibility, and its ability to annoy.

A central focus of Chapter 8 is the broader issue of the mobile telephone's effect on social cohesion (Putnam 2000). Will the mobile telephone contribute to or weaken the social capital? Clearly, the jury is still out. There is good evidence for both arguments. On the one hand, the mobile telephone can lead to "balkanization," in that we can escape our immediate situation and interact with only like-minded persons (Portes 1998). In the process, we do not just drop out, but we also colonize a part of the public sphere and reduce it slightly by our unwillingness to participate. Hardin's (1968) concept of the "tragedy of the commons" is seen here, in that the device provides an advantage to the individual while slightly diminishing the public sphere.

At the same time, the device allows for the development of stronger in-group ties. It leads to the sharing of experiences and emotions more immediately than almost any other mediated form of contact, save face-to-face interaction.

2

CHAPTER

Making Sense of Mobile Telephone Adoption

There are those who suggest that technology shapes society. There are also those who say the opposite: Society shapes technology. Finally, there are perspectives such as the affordances approach and the domestication approach, which seek a more nuanced — or perhaps more muddled — middle ground.

Our adoption and use of mobile telephony is a case study here. It has happened right before our eyes. On the one hand, the original engineers who developed the various systems are still among us. It is possible to talk to them and find out what possessed them to develop the system in the way they did. On the other hand, the users are here in spades. We need only go out onto the street to see the way that mobile communication is being used. Only a few years ago these people were not routinized users. They were fumbling through their first calls and they were trying to understand how they might respond to a text message.

Interaction Between Technology and Society

Mobile telephony is a technology that is quickly finding its niche. The technology has become reliable and easily accessible. In addition, it has been adopted on a large scale, and it is on the way to becoming a taken-for-granted part of the social landscape in many countries.

Nonetheless, we are still in the process of orienting ourselves and making sense of this technology. Many technologies have what one might call a crystallized solidity that makes reinterpretation difficult. One can, for example, think of cars and the massive system of roads, gas stations, marketing, and fabrication that is associated with them. While we can make small and almost inconsequential changes here, such as the style of the cars or the features available in the sound system, these changes represent only the small personalization of a massive system. There are, however, no substantial new areas of use that are being developed for the car-based transportation system.

On the other hand, relatively new technologies — such as mobile telephony — are more available for reinterpretation. The time, place, reasons for use, and way they are used are, in many ways, more open than in the case of the more thoroughly established technologies. Mobile telephony originally gained a foothold as a voice-based communication system. However, teens' adoption of text messaging has changed the nature of mobile communication. It has opened a new way of communicating and changed the way we orient ourselves to group coordination. Even as I write these lines, the mobile telephone is being transformed. It is becoming an anywhere/anytime Internet terminal, complete with the ability to send and receive music. Cameras are being added to terminals, allowing us to take, send, and receive pictures; in addition, "push-to-talk" features are being adopted. Location transmission services, where one can locate the physical position of friends and colleagues via the use of mobile telephones, promises to allow the same type of radical shift in the way we organize our daily activities. Thus, mobile telephony is moving away from its traditional base into new, uncharted waters.

The newness of the device means that to some degree we are making up the rules as we go along. We are finding new and unexpected uses for and ways of using mobile telephones. Let me illustrate this point: I was recently at an airport, waiting in line to pass through the metal detector on my way to a flight. I noticed a well-dressed fellow ahead of me talking earnestly on his mobile phone. As we came up to the maw of the X-ray machine, he continued talking on his phone while he removed his outer jacket, empted his pockets, and placed his briefcase on the conveyor belt. Finally, at the last moment and without hanging up, he put the phone on the conveyor belt and walked through the metal detector himself. On the other side of the barrier, he quickly recovered his phone as it came out of the X-ray machine and, almost without missing a beat, picked up the thread of his conversation. He refilled his pockets, picked up his jacket and briefcase while juggling the phone on his shoulder, and then disappeared around the corner en route to his flight. This illustrates how the mobile telephone pops up in new settings and, in some ways, recasts those settings.

From a sociological perspective, the process of socially defining the mobile telephone is revealing in itself. The rise of mobile telephony gives us the chance to observe the adoption of a new technology. Beyond providing insight into innovation, it affords us the chance to see how the innovation is accepted and how it causes the revision of existing values and practices. It allows us to see who is influencing the definition process and, in effect, whose toes get stepped on. The adoption of a mobile telephone means that we have to make adjustments and rethink how our "mental furniture" is arranged. The mobile telephone shifts ideas about where and when we can travel, how we organize our daily life, what constitutes public talk, and how we keep track of our social world. In addition, our use, or refusal to use, says something about us as individuals.

Technical/Social Determinism and Affordances

Before launching a more careful examination of the mobile telephone, it is useful to take a couple of steps back to consider the interaction between technology and society in general. Various notions of this interaction have been developed, including the technical deterministic, social deterministic, and "affordances" approaches. Finally, there is the line of thought that is more or less adopted here, namely, the domestication approach.

The technical deterministic view is well entrenched. It suggests that, in the final analysis, it is technologies that form and mold society. An encapsulation of the position is attributed to Marx in the statement "Give me a hand mill and I will give you feudalism; give me a steam mill and I will give you capitalism." In a no less technical deterministic statement, Mumford (1963, p. 14) notes that "the clock, not the steam engine, is the key machine of the modern industrial age." The core idea is that, when all is said and done, technical devices are at the root of social formations. This perspective can be read in Cottrells's social history of diesel locomotives in the United States (1945), Sharp's discussion of the introduction of steel into Australia (1952), and, in some respects, Eisenstein's examination of the printing press (1979).

The opposite perspective is the social determinism of technology, perhaps most thoroughly developed by Bijker and Law (Bijker and Law 1992; Bijker *et al.* 1987).[1] According to this approach, technologies are continually reinterpreted by users and given new, often unexpected, trajectories. Thus, while a mobile telephone is designed primarily as a communication device, it can conceivably function as a type of hammer, a shoehorn, a bottle opener, or even a type of flashlight, depending on the inventiveness of the user. While the primal use perhaps has, predominance, it is clearly possible to develop alternatives.

From the social deterministic perspective, we can see technology as a type of text that has an author (the designer) and is also "read" and interpreted by the user. Thus, the intentions of the author have a bearing on the interpretation, but the reader is also active in this process and can redefine the technologies as suited.

Both the technical and the social deterministic perspectives have their weaknesses. Technical determinism often assumes that the technology sprang, as it were, fully formed out of the head of Zeus. There is inadequate attention paid to the idea that the technology was created in a specific social context. In this way, there is no clear idea of what constitutes the social and what constitutes the technical (Hutchby 2001). The opposite critique applies to the social deterministic view. If we carry this position to its logical end, technologies have no intrinsic qualities. A particular technology, be it a pillow, a screwdriver, or a rocket ship, can be reinterpreted. In this way, their identity is more or less based on their socially negotiated profile.

A critique of both positions is that while they stake out positions as to the nature of technology, they seemingly operate at such a high level of abstraction that there is really no way of proving or disproving the assertions they make. Given this, we are left to interpret them as ideological positions used to guide our inquiries. We can unkindly suggest that both the technological and the social deterministic positions are left arguing as to how many angels can dance on the head of a pin.

A type of middle position has been developed in the idea of affordances. The original statement of the theory was provided by Gibsen (1979) and further developed by Norman (1990). The affordances approach describes how the physical characteristics of an object interplay with the way in which we perceive and interpret the use of the object. Gaver suggests that affordances are characteristics that are more or less directly available to the perceiver. Gaver, for example, examines how a vertical door handle "affords" pulling in the mind of the user and on-screen buttons that seem to protrude from the PC screen "afford" pushing (1991). The core idea is that "the properties of objects determine the possibilities for action" (Sellen and Harper 2002, p. 17). The idea is further applied to interactions with technology by Hutchby in his analysis of conversation and technology (2001) and by Sellen and Harper in their analysis of paper (2002). Not surprisingly, this approach has been most completely elaborated in the area of design.

The idea is useful in that it highlights the shortcomings of both the technical and the social deterministic perspectives. It points to interaction between the physical and the social in the imbedding of artifacts into our lives. On the negative side, however, the notion of affordances also suffers from several drawbacks. The approach often focuses on the design of the technical object and does not seem to pay attention to issues such as culture, age, and experience of the user to the same degree. Following this thought, you might suggest that a mouse attached to a PC "affords" the possibility to move the cursor about the screen. This is taken as a matter of course for many millions of PC users every day, and as such it can be interpreted as an affordance. Indeed this is seen as the genius of the mouse as a way of manipulating activities on the PC.

However, use of the mouse is a learned behavior. One of the first times my 3-year-old daughter used our PC, I sat beside her as she doodled with a drawing program. At one point, the cursor had been moved from her picture to the menus under the drawing area. She clicked and became frustrated that the drawing function of the cursor was replaced with a button-pushing function. In order to continue drawing, she had to move the cursor up into the drawing area. I told her to "lift it up" or words to that effect. I was, of course, referring to the cursor on the screen. However, she interpreted my instructions differently and physically lifted the mouse up from the desk, anticipating that the problem would be solved. Clearly, I was referring to what was, for me, a taken-for-granted relationship between moving the mouse in some direction and the resulting movement of the cursor. Her unschooled approach led her to tackle the problem with a perfectly logical, but incorrect, solution. The situation, however, points out how the assumption that objects simply radiate functionality is off base. Rather, the objects have to be placed into a larger context and understood from that perspective.

Applying the same point to mobile telephony, we can see the difficulty of the affordances approach in describing the broader social motivations for the adoption of text messaging. The menu structure and form of text entry augured against the use of texting, since it was — and indeed still is — difficult to enter into the system and compose messages. The success of SMS results from determined individuals who persevered in their desire to communicate. To simply call this an affordance seems to stop short. Rather it is useful to look at the broader social context. What is it about the teens who first adopted the system? What enticed them to explore this avenue of communication? What were the social pushes and pulls that eventually resulted in what has become a culture of SMS? How has SMS changed the dynamics of youth culture, and how has youth culture changed and redefined SMS?

While it is a slightly unfair criticism, one comes away with the sense that the affordances approach focuses on the person–machine interaction "then and there." It is at its best when describing design issues and perhaps weaker when considering the broader social context. The point here is that the broader sociocultural context is also of interest. Indeed one needs to understand issues at this level in order to understand the design of technology.

Another, perhaps more serious, problem with the affordances approach is that it is tautological. In many respects, the concept of affordances suffers from the same critique that has been leveled at the structural functionalism of Talcott Parsons. It has been said of structural functionalism that if a social function exists, there must be some functional reason for its existence. Thus, the functions are defined in terms of their functions. The argument is circular, and the discussion is tautological (Turner 1986, p. 49).

There is often the same sense in the discussion of affordances. If an object is used in a certain way, then it has an affordance for that use. In a sense, $A = A$. There is no analytical leverage in this equation. Any use to which an object is put can be interpreted as a naturally arising characteristic of our interaction with that object.[2] There is no sense, for example, that the statement is falsifiable or, for example, that there can be degrees of affordance. Thus, the concept loses much of its analytical power.

Domestication of Information and Communication Technologies (ICTs)

The approach that has been most instructive in the work presented here is the so-called domestication perspective. In many ways, it is a compromise between technical and social determinism, while at the same time it avoids the narrowness of the affordances position. Following on the work of Douglas and Isherwood (1979), this line of thought was developed by several researchers, most particularly Roger Silverstone and Leslie Haddon, in the United Kingdom (Silverstone and Haddon 1996; Silverstone 1994; Silverstone *et al.* 1992). The approach was originally developed to examine the adoption of technologies in the home, but has been expanded to include mobile technologies (Haddon 2001). It has also been used extensively in Scandinavia (Sørensen and Østby 1995; Vestby 1996).

Haddon notes that there are several general points characterizing the domestication approach (2001). The first is that the emphasis is on the consumption, not simply the purchase, of an item, be it a material or a nonmaterial artifact. The point is that in order to understand the role of an item in a person's life, a researcher must have an overview of the negotiations and interactions associated with its acquisition and its ongoing role in the home or the social group. The second point — a point that is central — is that adoption should be viewed as a process. We go through a whole series of negations (both with ourself and with others in the home or social group) before deciding to acquire a particular artifact or service. These negotiations can involve the degree to which we really need the object, the amount to be spent, the conditions in which we should use it, its placement — physically and temporally — in our lives and in our homes, etc. These negotiations can indeed take on a moral tone when involving the adoption of technologies by teens. As we will see later, teens' adoption of the mobile telephone involves not just the simple purchase of an object, but the touchy interaction between parents and teens in the emancipation process.

The third point outlined by Haddon is that domestication is "not a one-off process." Rather, even after an artifact has been brought into our social or domestic sphere there may be ongoing discussions/negotiations/arguments regarding its role. For example, the placement, time commitment, programming choice, or volume level of the TV can also lead to interactions in which household members feel the need to refine the role of the device and establish new rules as to how the object may or may not be used. It is also noted that the domestication of an information and communication technology (ICT) is never completely successful.[3] Household members can carry ambivalent feelings about a device, such as the availability afforded by mobile telephones or the time consumption and blurring of work/private life associated with the PC.

A fourth point—one that distinguishes the domestication position from that of affordances — is that domestication is not only a mental process carried out by an individual but also a social interaction between individuals. There can be gatekeepers, as in the case of parents with teens' adoption of the mobile telephone. There are also influential individuals who perhaps help novice users understand how the device should be used, displayed, and generally integrated into their lives. In addition, the ownership and use of various artifacts helps us to pigeonhole others. Thus, our own consumption becomes a part of our own social identity. Further, others' consumption is a type of lens through which we see them and through which we interpret their social position. Finally, and correctly, Haddon points out that "the role that ICTs come to play and their meaning for us both is shaped by the rest of our lives and can be shaping in their consequences. In other words, how we experience them is not totally predetermined by technological functionality or public representations but is also structured by social life" (Haddon 2001, p. 5). Domestication then looks at the acquisition, display, function, and consumption of particular objects (or even services).

Silverstone *et al.*, for example, point out that devices such as the mobile telephone are doubly articulated (1992, p. 21). That is, the mobile telephone is not only a physical object, indeed an object to which we carefully assign meaning, but also a medium through which we communicate and through which we maintain social contact. Considering the former element here, we can see the mobile telephone as a type of jewelry. In some cases, this is a peripheral design issue. In other cases, the display of the object is central.[4] As with jewelry, we must learn how to appropriately display it in social situations and use it in a socially graceful way. In addition, it is an object that is assigned a role in our sense of fashion and being fashionable. As we will see in Chapter 4, it is a way in which we assert our identity and it is a way in which others interpret our identity as wearer/user. Speaking of adopting the PC into the home, Silverstone *et al.* (1992, p. 23) note that

> *All technologies have the potential to be appropriated into an aesthetic environment (and all environments have, in some sense, an aesthetic). And many are purchased as much for their appearance and their compatibility with the dominant aesthetic rationality of the home as for their functional significance.*

Undeniably, the artifact has its functional features, but it also has its social profile. One person's sense of how to elegantly display and use the mobile telephone can easily be the bane of another.

The domestication approach describes several steps in the adoption cycle (Silverstone and Haddon 1996; Silverstone 1994; Silverstone *et al.* 1992). These include *imagination, appropriation, objectification, incorporation,* and *conversion*. These stations describe the movement from having the idea that an object or service would be a useful addition to our life to the purchase of the object and its imbedding in our life. Finally, the process describes how an object becomes externalized as part of our social profile. Think about the elements of the process a little more carefully. *Imagination* is the way in which a device, such as the mobile telephone, a new type of scarf, or any other innovation enters our consciousness. *Appropriation* describes that portion of the consumption process in which a particular object leaves the commercial world and enters our sphere of objects. This stage in the cycle includes the sense that we know of the particular object or service and, further, have the sense that it could somehow fit into our life. Friends or advertising may be the link through which we understand the existence and the value of the object.[5]

Objectification describes how a particular object or service comes to play out our values and sense of aesthetic. This stage in the domestication process describes how we think through the way in which an object will fit into our world. At issue is what its placement, use, accessibility, time consumption, etc. say about us. The sense of *objectification* is that we somehow see that we are working out our identity through the ownership and consumption of a particular artifact. That is, we are involved in making manifest our sense of identity through an array of objects or services that are perhaps selectively used to engender a particular effect. Nonmaterial objects, such as the ringing sound on our telephone, can also be seen in this light.

In a slight amendment to the work of the domestication theorists, I choose also to draw on the work of Goffman in this context. Indeed his work (1959, 1963, 1967) as well as that of Meyrowitz, who develops the notion of technologies as social actors (1985), plays a central role here. Goffman's concern with the staging of everyday life, his focus on the ruses and props used to achieve an effect, and the way in which we strive to maintain a façade are all useful tools in the analysis of mobile telephony.

Goffman's concern with the construction of everyday life provides insight into how we approach and appropriate ICTs. His work allows us to see how these technologies, and in particular a technology so intensely personal as the mobile telephone, are used in the development and maintenance of a façade. In addition, his work includes the notion that there are frontstage and backstage dimensions to the use of mobile telephony, both in terms of our colocated situation and in terms of our telephonic interlocutor.

Returning to the domestication approach. While *objectification* focuses to a certain degree on the aesthetic, *incorporation* directs itself more toward the functional. In effect, they are two sides to the same coin. *Objectification* is the way in which an artifact comes to crystallize a sense of self. *Incorporation* describes the functions of these artifacts, not just those described in the ownership manual, but all of those that are applied by the users. *Incorporation* is also concerned with the temporal assimilation of the objects into time structures and routines. In addition, this point in the cycle examines how a particular device is incorporated into an existing array of artifacts.

A central point here is that the functions as well as the functionality of an object can shift and change over time. Indeed technologies can have extended and varied careers in the home as they move from being the "latest thing" to being forgotten dust traps (Haddon and Silverstone 1993). Writing on the heels of what we might call the "Sinclair PC revolution" in the United Kingdom, Silverstone and the group around him were well aware of how the original bloom associated with PCs had wilted. The PCs had represented a significant investment as well as participation in the brave world of the information revolution. By the 1990s, the PC had been slowly redefined and pushed more and more into the periphery of the domestic sphere. Thus, in their descriptions of ICTs, Silverstone, Haddon, and the others had a clear idea as to how the original notions about the ways PCs would be used had shifted and changed as users became aware of their possibilities and limitations.

The final point in the cycle is *conversion*. According to Silverstone *et al.*, it is in this phase of the cycle that others incorporate their understanding of the artifacts in their broader understanding of the person consuming the artifact. It is at this point when the person who purchased and is using the item hopes to realize its social effect. Obviously, however, effect is in the eye of the beholder. While we can consume artifacts with one image in mind, others are free to interpret the situation from their perspective. Drawing on Veblin, Silverstone *et al.* note that "the *appropriation* of an object is of no public consequence unless it is displayed symbolically as well as materially, for through that display a household's (or a household member's) criteria of judgment and taste, as well the strength of his or her material resources, will be asserted and confirmed" (Silverstone *et al.* 1992, p. 26).

When we progress through this cycle, we start with an understanding that a particular artifact exists first as an imagined consumption, as in a type of extended window shopping. Based on this, we perhaps decide to incorporate the item into our daily life. Its purchase removes it from the commercial world. Subsequent to purchase, we must work through how the item will be arrayed and also explore its actual use. Here we go through the real consumption and thereby examine the qualities of the item and map them onto our imaged self. The final turn in the cycle is that the object becomes an element in others' estimation of us. One is reminded here of Berger and Luckmann's process of institutionalization, in which a behavior or object moves from its social construction into a type of taken-for-granted institutionalized status (1967).

Before fully embracing the domestication approach, there are several points to be made. First, domestication is generally a micro-level approach. It is focused on the everyday life of the individual in a particular context. There is not the sense that it is a part of, for example, the technical deterministic project to clarify technology adoption on a broader, macro scale.

Another caveat is that we need to be careful when assuming that domestication is a sequential approach to adoption and consumption. Indeed, Haddon overtly recognizes this when noting that domestication is not a simple "one-off" process (Haddon 2001). While there are linear elements to domestication, i.e., the introduction of the artifact, its purchase, and finally its being an imbedded part of the presentation of self, it is not necessarily the case that all of the stages are entered sequentially. People seem quite agile in the ability to mentally objectify items long before the actual purchase has taken place. We can see this in the text of a group interview in which Frank[6]—a mid-20s salesperson—describes the type of mobile telephone he wants.

> ***Frank:*** *I want a mobile telephone that you can feel in your pocket, so that will cost a little extra; otherwise I don't want to bother with one.*
>
> ***Marit:*** *Is that anything to brag about, that it's expensive? Isn't it important that we can call cheaply and that we...*
>
> ***Frank:*** *I am happiest by having the best. The best isn't good enough.*
>
> ***Marit:*** *I don't understand that, but...*
>
> ***Frank:*** *All my friends are great, they have the best cars and the best and most expensive stereos, the most expensive TVs, and the most expensive mobile telephones. And since nobody has bought that Philips yet, then I will be the first.*

Frank describes the feel of the yet-to-be-purchased device in his pocket. He has, perhaps, been allowed to finger the mobile telephones of his friends and thus has a sense of their physical properties. In addition, we sense from his comments that he yearns to buy a particular type of mobile phone. While he is actually in the prepurchase *imagination* phase, we can see that he is, in some respects, quite far into the *objectification* and even into the *conversion* phase as described by Silverstone *et al.* That is, his intended purchase has almost become an externalized part of his identity. The fact that he has no mobile telephone is almost irrelevant. He is well oriented in the market. He knows the brands and styles. He knows what type of phone he wants and the image he wants to project. In many respects all the elements are in place; the only thing missing is the mobile telephone itself.

By way of contrast, there are those who actually have a mobile telephone but feel somewhat uncomfortable with it. In this case, neither *incorporation* nor *conversion* is completely accomplished.

> ***Irene (42-year-old single mother of two):*** *I have one but I don't use it much. I saw I just got a bill and I had calls for 24 kroner (approximately $3) or something like that during the last three months. I just have it; it is for those times that I needed it, like if I am sitting in traffic, if the children are alone home .... That is why I bought it, if, for example, they should be alone—for example, I am at a parent's meeting at school or something like that, you know. Then I can have it with me, so they know that they can reach me if there is something.*

Irene's comments contrast with those of Frank. Where Frank seems to have a secure sense of the mobile telephone's role in his life and of the image he wishes to project via the device, we are left with the sense that Irene is grasping. While she is in fact the owner of a mobile telephone, she has not really integrated it into her sense of self to nearly the same degree as the brash fellow in the first citation. The point then is that the sequence of domestication is not necessarily a linear progression that proceeds through the purchase and the imbedding of a technology into our life.

Another point of critique is that while domestication can be applied to the process of adopting and consuming novel objects, it loses some of its power if we apply it to items that are more mundane. As soon as an object becomes a routinized part of everyday life, much of the analytical power of the approach disappears. It is clear that the domestication approach is useful in describing the adoption of ICTs as well as other items that are prominent and are being adopted. We can think, for example of sport utility vehicles in this context. It is good at describing how they elbow their way into our lives and, in effect, make room for themselves. However, it is more difficult to think of domestication as a way to describe the purchase of, say, salt or toilet paper. In this case, the deliberation process is relatively curtailed, and while the purchase of these items may have something to say about our façade, these statements are relatively meager when compared to the more ideologically demanding issues surrounding the purchase of a mobile telephone. Obviously, mobile telephony will follow other consumer items into the gray future where they have lost their valence. To the degree this happens, domestication "theory" loses some of its ability to describe the situation. Finally, domestication is not a theory. That is, it does not seem to provide us with provable hypotheses. Rather, it is more of a method or an approach to research.

This said, domestication as an analytical perspective has several advantages over the alternatives noted earlier. When we think of the technical/social deterministic approaches, domestication has the advantage of not necessarily staking out a final position. It assumes we arrange our lives and define ourselves vis-à-vis technical objects and that technical devices have consequences for the arrangement of our everyday lives. It limits itself, however, to the sense that objects and artifacts are, from the perspective of the individual, preexisting and that the individual is not solely responsible for their interpretation. In addition, it has no trouble accommodating the notion that social factors are important when understanding the use of technology.

In effect, it recognizes both positions, and thus it avoids the need to say that one or the other is predominant. In this way, it is a pragmatic approach. It accepts the idea that technologies have effects on the organization of society and the notion that society forms technologies. The point is not to follow either the technical or the social deterministic approach to the bitter end, but rather to see how people make sense of the situation in which they find themselves.

Finally, domestication avoids the narrowness of the affordances approach. Domestication looks at both the interaction between the individual and the artifact and the social context in which the artifacts are being defined and used. In addition, unlike the affordances approach, domestication looks at the role of the particular item in the way that life is lived out through our consumption and the use of various objects and services. It also treats the adoption and use of objects and services as dynamic and changing.

In line with the domestication approach, the work here is not focused on the testing of hypotheses but rather on the description of an ongoing process. While the work may be able to make comments on the interaction between technology and society, the central focus is the description of the ongoing interaction between the mobile telephone and its accommodation in society.

Methods and Data Sources

As one can guess from the preceding discussion, the approach used here is based on empirical studies—both qualitative and quantitative—of mobile telephone adoption and use. These include a broad range of survey data, personal and group interviews, and observations.

As a point of departure, I use the recent experience of Scandinavia, specifically Norway, to examine this phenomenon. However, the material in this book is not limited to data from Norway or even Scandinavia. Rather, empirical analyses and observations carried out by other researchers from Europe, the United States, and Asia are included where available. These include qualitative interviews, observation, or contextual studies and quantitative analysis.

The qualitative material includes interviews with several hundred persons from a broad spectrum of social situations in Norway over the past 5 years. Interviews have been carried out with teens, young adults, parents, and elderly persons, in group situations, individually, and in the home. In addition, the material includes qualitative material gathered in the context of European-wide studies of ICT use. The most extensive of these was a series of 36 group interviews across six countries carried out in 1999. Finally, taking a page from Goffman, observational studies have been used in the preparation of this book. This has been a particularly important element in the analysis of mobile telephony as a disturbing influence, as discussed in Chapter 6.

Beyond the qualitative material, the book includes the analysis of a large body of quantitative databases, including almost 30,000 interviews in various surveys gathered in a variety of situations in both Scandinavia and Europe. The quantitative material includes annual surveys of teen mobile telephone use between 1997 and 2002. These constitute roughly 7000 telephone interviews from this period. In addition, the quantitative material includes material from the Norwegian media-use survey from 1999 to 2002, including approximately 5000 additional surveys covering all age groups, and from the Norwegian time-use survey for 2000 that included approximately 3700 individuals.

Beyond the strictly Norwegian material, I have also participated in the development and execution of European-wide studies. One was a survey of more than 9000 individuals from nine European countries under the auspices of the EURESCOM P-903 project. All age groups and sociodemographic groups were included in this analysis. Another survey is the EU's e-living project, that is a panel survey currently taking place in five European countries. The first wave of data analysis was completed in 2001 and the second in 2002. A complete overview of the surveys and studies used is shown in the Appendix.

3

CHAPTER

Safety and Security

Introduction

During the early and mid-1990s the mobile telephone was often associated with yuppies. One can recall Michael Douglas, playing the role of the unethical financier Gordon Gekko, in the film *Wall Street.* Gekko used a mobile telephone from the beach outside his luxury home when calling to Bud Fox to deliver his "wakeup call." This use of the device played on the image of the mobile telephone as a form of conspicuous consumption available only to the wealthy.

As though to illustrate Haddon's suggestion that our take on technologies is "not a one-off process," the widespread adoption of the mobile telephone has replaced this image in the popular imagination (Haddon 2001). While still being seen as relatively expensive, one of the most common popular images is that it provides us with a form of security—that is, the sense of being out of harm's way—and safety—in which one is indeed free from harm. To be sure, the notion of a mobile telephone as a lifeline is one of the central images of the device.

Qualitative interview data as well as quantitative analysis indicate that mobile telephony provides a safety link for those who have chronic sickness as well as those who find themselves in dicey situations. Here's an illustration: One American reported the following rather dramatic situation.

> *After lying beside a car in the dirt while bullets whizzed past my head, my friend called 911[1] on his cell phone. By the time we had crawled into our vehicle and driven toward the canyon rim, the police had arrived and had the two suspects in hand. They had been shooting at a paper target propped against sagebrush on the rim of the canyon, their bullets dropping into the canyon on unexpecting residents.[2]*

This dimension of mobile telephony has evolved into a general sense that the mobile telephone provides us with security. Many people who have never been in a genuine emergency carry mobile telephones "just in case" something happens.[3] In the words of one Norwegian informant:

I got my first mobile telephone eight or nine years ago. I bought it simply because I often go to the mountains alone. Suddenly your car stops and it is about –5 farenheit. What am I to do then? ... Plus I have a cabin without electricity.... It is insecure, and things can happen, suddenly break a leg, the kids might need me, and so on. (Kjell)

Other informants spoke in terms of security combined with geographical mobility. They talked about freedom of movement in relatively uninhabited forests and wilderness areas.

I thought that when I was retired I will go walking exactly when I want to, especially in the mountains.... Then there is the problem that my wife says that she doesn't want to take the chance of me being up there. You go skiing and can be laying there and freeze to death... If I go skiing and I am lying there I can call Norwegian Air Ambulance [A helocopter rescue service], I am a member, then I can lay there and wait for help. (Karl, retiree)

These comments underscore people's sense that with the mobile telephone we are never alone in an emergency, or at least we are more available than before the rise of mobile communications. Indeed there are some rather striking examples where mobile telephony has come to the assistance of people in difficult straits. People in London started rescue operations for friends stranded on a boat off Indonesia after receiving a text message. A woman who became stranded in the outback of Australia was able to alert her bother to her plight. He was able to assist in rescuing her. A British skier lost on a mountain in Italy was able to send an SMS to his mother, who lived in Derbyshire in the United Kingdom. In another case, two British climbers who, due to a blizzard, became stranded for three days on mountain in Switzerland were able to notify a friend in London via SMS. The friend notified local authorities, who sent a helicopter to rescue the climbing group.[4]

This argument can be taken to its logical extreme, as during major catastrophes such as the terrorist attacks on of September 11 in 2001. In that case mobile communication devices were used to coordinate and organize activities as part of the rescue efforts, and also as a way for victims and their families to exchange comfort in the face of the unimaginable (Dutton 2003; Palen 2002; Rice and Katz 2003).

All of this does not ignore the fact that the mobile telephone can also enable illegal activities such as drug dealing, gang fights (Lien and Haaland 1998), and even the terrorist attacks themselves by allowing the terrorists to coordinate their activities (Dutton 2003) and reportedly detonate explosive devices remotely, as in the case of the attacks in Bali and Jakarta. At a far less dramatic level there is also evidence that the device can be the cause of accidents, particularly while driving (NHSTA 1997; Redelmeier and Tibshirani 1997; Cain and Burris 1999; Recarte and Nunes 2000; Strayer *et al.* 2003; Strayer and Johnston 2001). Thus, as with so many other things, the mobile telephone has both advantages and drawbacks. On the one hand it can provide for safety and promote a sense of security, at the same time it can facilitate malicious events.

The Mobile Telephone as a Contribution to Security

Safety and security are among the most basic reasons to own a mobile telephone. In a survey of European users,[5] respondents were asked to what degree they agreed or disagreed with the statement "The mobile telephone is useful in an emergency." We found that approximately 82% of the respondents were in complete agreement. There was no other attitudinal indicator with regard to mobile telephony that had such an extreme score.[6] There were no real gender differences in the results.

Approximately a year after the broader European analysis, a similar battery of questions was included in a study done in Norway. Instead of simply saying that the mobile telephone is important in an emergency situation, the wording was changed to say that "A mobile telephone is *most* useful in emergency situations" (Figure 3.1). The slightly stronger statement provided some interesting insight. In this case, the those who completely agreed fell to 56%. Further, analysis of the latter statement in the Norwegian study shows that it was women[7] and the elderly[8] who were likely to agree with the idea that mobile telephones were most useful in emergencies. Where only about half of the men agreed completely, 64% of the women had the same attitude. The differences were even more striking in terms of age. The data shows that 77% of those over 67 years of age were in complete agreement with the idea. Only 44% of those in the 35–44 age category had the same attitude. Interestingly, the youngest users were more likely to agree with the statement than those who were of middle age.

There is a slight irony here. While the elderly are the most likely to see the mobile telephone as a disturbing influence in the public sphere, this group also

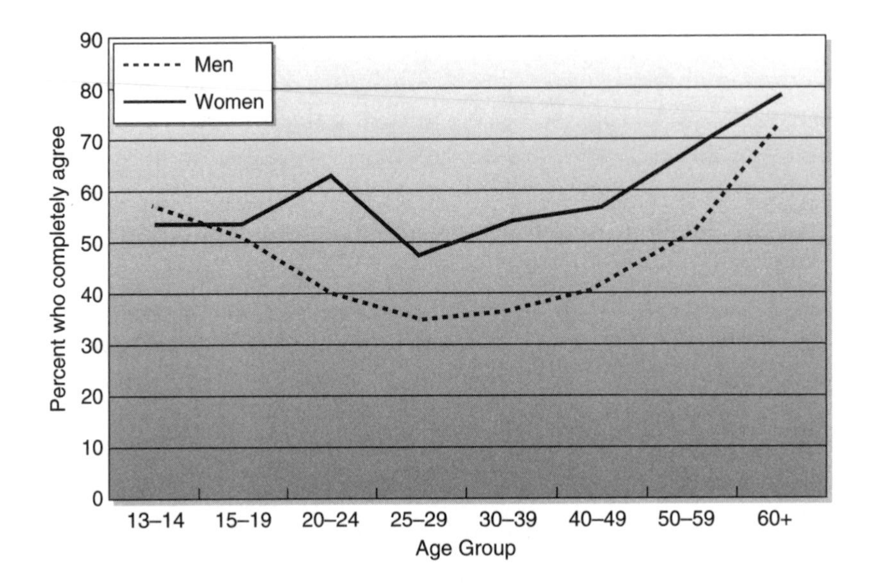

FIGURE 3.1 Percent of persons who completely agree with the statement "A mobile telephone is most useful in an emergency situation." Norway, 2001.

has come to terms with the technology as a type of helping hand should the situation turn against them.

Material from the United States and Australia points in the same direction. In a survey of persons in the United States reported on by Wrolstad, location services and emergency response via mobile telephone were among the most attractive functions of the mobile telephone. This was particularly true among the elderly (2002). The emergency use of mobile telephones is also seen as an essential dimension of the technology in Australia (Chapman and Schofield 1998).[9] Thus, while the popular discussion of mobile telephony often focuses on other dimensions of the phenomenon, these findings suggest that safety and security are the basic functions.

Use in Situations Where There Is a Chronic or Acute Need for Contact

The mobile telephone has an obvious role in the lives of those in chronic need of help. For those persons, the ability to contact others regardless of where they might be is a boon. The same can be said for persons who experience an acute need for help. This aspect of mobile telephony has been abstracted into the sense that access to the mobile telephone has become a way to ensure our safety in a variety of situations.

The mobile telephone has developed a role as a type of lifeline for people with a chronic need to contact emergency services.

> *[I use it] because of my health, I have it with me when I am not at home. If I am sick, I can call an emergency number or something because I have terrible asthma and I can have an attack that I have to take medicine, to say it that way. And so, it is good to have [the mobile phone] with me. It is a comfort to have it, I feel. (Terje)*

> *There is a lot of security [with a mobile telephone]. My wife is handicapped, so she needs to be able to call me and I need to be able to call back again and tell her where I am and what I am up to. So I am very positive about the mobile telephone, but I don't use it other than when I am away so that she can reach me. (Alf)*

> *I have a mother-in-law that bought a mobile telephone. She had angina and they have a cabin in Bodø and she bought one because she though that it was better to have a mobile telephone. (Lilian)*

In each case, the individuals noted a chronic medical condition and the potential need to contact emergency services.

People with physical impairments use the mobile telephone to allow themselves a certain emancipation. Palen *et al.* report on a person who while wheelchair bound carried a mobile phone when working in the area surrounding his home in case he fell out of the chair and needed assistance (2001). Foreshadowing the discussion in the following chapter, he also used the device as a coordination tool. He reported that he contacted others with whom he was planning to meet when physical barriers hindered him. Thus, while there was a safety dimension to this person's use of mobile telephony, his use also included dimensions of coordination and emancipation.

Research has shown that the use of emergency landline phones increases the possibility of survival in the case of cardiac arrest (Chapman and Schofield 1998). Clearly, mobile telephony extends this potential and can reduce the time needed to respond to medical emergencies when not near a landline telephone (ComCare 2002; Sakarya 2002). Mobile communication adds a new dimension to this situation (Gow 2002). With respect to health-related use of the mobile telephone, it is reported by Potts that use of the system has significantly reduced the time needed to notify emergency services (2000). In addition, there are various initiatives related to the so-called enhanced emergency number services that would allow location transmission and thus more precise location of those who are making mobile emergency calls.[10] It has been estimated that the number of wireless emergency calls in the United States has gone from less than 10,000 per day in 1988 to almost 120,000 as of 2000 (ComCare 2002). We can expect this trend to continue as the adoption of mobile communication continues.

People faced with chronic conditions often showed great ingenuity in the way they exploited mobile communication technologies. In the case of a 50-year-old mother, Bjørg, the telephone was used as an aid in her role as the primary caregiver. Bjørg needed to be continually available because her child suffered from irregular seizures and the health of her parents had begun to fail. Given this situation, Bjørg ensured her continual availability via a range of devices and services.

> *We have a car phone and a little portable. Most often, when I am out of the house here I activate call-forwarding. We have a problem because we never know when the school can call, and I have parents that are sick. We never know when something can happen. And so it is just transferring our home number [to the mobile telephone].... We need to get that information that is necessary for family members because it can happen whenever. So I feel more secure that they can reach me.*

In consideration of her parents, she used both landline and mobile telephony. While there was a health-related dimension regarding her parents, Bjørg was also concerned with ensuring their peace of mind.

> *When it comes to my parents, they know that they can reach me if things happen. And in the car, it is ok because we drive down to Sandefjord, where they live, and down through Vestfold there are so many [car] accidents. And if we don't come at about the correct time they are very insecure and afraid. Then we can just call and say that we are in bumper-to-bumper traffic and they can relax. And they can reach us regardless because they have the number.*

Bjørg describes her use of the technology not only to be available in case she is needed by her parents, but also to assuage her parent's fears while she is driving a particularly dangerous route, and also ensuring her perpetual availability.

Bjørg's situation meant that she was prepared to go to some length to be available to her family. In the case of her summer cottage, mobile telephony was unavailable and so she had developed a rather extended chain of technology through which she could maintain contact with her parents.

Bjørg: *The summer cottage is such that there is no mobile coverage there . . . . But that is no problem because the house next door has a telephone and so we transfer our telephone to there.*

Interviewer: *Ok, so they can get a message?*

Bjørg: *Yeah, that is no problem. We take a cordless [telephone] with intercom, so the people in the house can just push the button and so we know.*

Bjørg seemed to be a master of cobbling together wireless communication systems and the strategic use of call-forwarding. She understood how to shift calls between landline and mobile systems to allow her freedom of movement while nonetheless being on call for her family should a situation arise. She was able to develop ad hoc communications networks that allowed her to be continually available for her parents and her child should the need arise. In addition, she was allowed a relatively wide range of action herself.

The mobile telephone helped the informants cited here deal with a constantly threatening medical situation. Another application of mobile communication is the use of the device in acute situations, such as accidents and medical emergencies. This is relatively common. Chapman and Schofield noted that more than 12% of mobile telephone users in Australia had reported an accident involving others and 6% had used the mobile telephone in a medical emergency (1998). Material from the Norwegian interviews included comments about the mobile telephone being used to summon help.

Things can happen. I have just experienced that a friend of mine was in an accident, and if it hadn't been for a mobile telephone at the site of the accident it might of been worse. (Mina)

I was involved in a collision and I had a mobile telephone. It was good to have that. It was out in the country and [it was] a long way to the next house. It was really good to have it. (Harald)

We once had a camping trailer and we ran over a girl and it was good to have a mobile telephone . . . . It was good for calling in assistance. (Rune)

These informants provide us with concrete examples of how they used the mobile telephone in emergency situations.[11] The ability to summon help to remote locations was seen as a positive dimension of the technology; indeed, Lemish and Cohen report the same based on their work in Israel (2003).

Abstraction of Security vis-à-vis Mobile Telephony

While it is common to hear of either actual chronic or acute situations in which the mobile telephone is used, it is far more common to hear discussions of its potential use. In the words of 17-year-old Andrew, "It is good to have if something happens." Indeed the purchase of a mobile telephone as a way to enhance security was a widespread sentiment in the pan-European qualitative study reported by Klamer *et al.* (2000). Informants in The Czech Republic, Spain, and France all emphasized the role of the mobile telephone. As further confirmed by the quantitative material cited earlier, this was particularly true for women. It was found that notions of security were often used as a justification for the initial purchase of a mobile telephone. This sentiment has become almost a formulaic version of why a person needs to have a mobile telephone. We see it, for example, in the urban legends of hapless hikers calling from treetops while a bear prowls below (Håberg 1997).

Some informants had an almost palpable excitement when discussing the geographical liberation provided by the mobile telephone:

> *Freedom of movement, a feeling of security, others have, perhaps, elderly parents that they don't dare leave behind ... Mobile telephones are helping to create a new type of freedom that we did not have before. You had to be there, not just with your job but also with the family. [Now] the family can communicate over undreamed-of distances (Marta).*

It is common for informants to express the sense that the mobile telephone was a type of insurance. There was not the need to express specifics; rather, the mobile telephone was seen as a bulwark against potential accidents and mishaps, particularly when you are alone. Eighteen-year-old Nora also brings up this perspective. "I think more about security if you are at a summer cottage, for example, and something happens. Then it is comforting to have a mobile phone, at any rate if you are alone." Along the same lines, Einar comments, "If somebody gets hurt or if there is an accident or if you come someplace and don't know where you are or where you are going, you can call." Seventeen-year-old Andrew says, "If you end up someplace where there is no phone and you cannot get the last tram [at night] or something like that, then you can call and say where you are and get out of there or call a taxi or something. You have a chance."[12]

As we have already seen in the case of Bjørg calling her parents, the mobile telephone was also seen as a way to alleviate others' concern. We also see this in the comments of Grethe, a woman with grown children who live in another part of the country: "When we are driving over the mountains, for example, we have a very long way to drive over the mountains and we always have the mobile with us. We tell the children before we leave and how long [we expect to be]. So if we get stuck or if something happens, then the mobile is great. We use it only for that." Another woman, Ida, comments on the same general use of the device:

> *[Mobile telephones are] great at the cabin. We travel back and forth and the whole family is not with us. Then it is good to know that the trip has gone ok and that you have made it where you are going and on time, calling to check up a little. I think that is a good idea. Especially when one is going from one place to another, it is good to have a mobile telephone. (Ida)*

Along the same lines, interview material showed examples of what Rakow and Navarro called "remote mothering" (1993). In one specific case a single mother, Marta, used the mobile telephone when circumstances meant that she had to be away from her children in the evening.

> ***Marta:*** *I remember that I was scorned one time because the mobile telephone we had was a really big one and we always had it when we were out*º*. I always had that big suitcase with me because that was my best babysitter. But I stand by that today we have a secure situation because of these mobile phones.*
>
> **Moderator:** *How is that?*
>
> ***Marta:*** *It was my babysitter. They [her children] were home alone while I was out doing other activities and then I had a mobile telephone with me, that big one*º*. I worked in a church, evening work, I remember, and then I had that big mobile telephone with me and the children were at home alone when I had to work. And that was my babysitter; they could call to me if something happened or if they were afraid or I could call them. I was often the one who called home, but they called me if they needed to say something or if they were afraid, and then I could come home if I needed to.*

The comments here refer to the potential use of the mobile telephone as a safeguard against potentially precarious situations. Generally, no emergencies arose. However, the comments show how they relied on the mobile telephone in various situations as a latent security link.

Another cross-generational example of mobile telephone use is seen in the example of Reidun, an elderly woman who lived alone. While many of the respondents took the responsibility to get their own mobile telephone, others had received one from their children and grandchildren.[13] The woman's experience, however, shows that the technology was not completely integrated into her everyday life, and thus the device did not seem to fulfill its potential role in the provision of security.

> ***Reidun:*** *I have been opposed to mobile telephones because I think there have been too many of them. But on my 75th birthday, my children and grandchildren gave one to me even though I wasn't so excited about it. I didn't know what I was going to use it for.*
>
> ***Alf:*** *But now you like it?*
>
> ***Reidun:*** *No, not so much. I have it with me when I am out driving and I have it with me at the cabin since I thought that I could be at the cabin alone since I thought that that would be very nice. So I was there one night alone and I didn't like to be alone. But I don't use it much at all.*
>
> ***Alf:*** *But on that one night you felt secure?*
>
> ***Reidun:*** *Yeah, you can say that*
>
> ***Moderator:*** *When was the last time you used it?*
>
> ***Reidun:*** *I don't know, it was a long time ago. I turn it on every once in a while. But there are so few that call, so I don't have much use for it. But it is security, if something should happen.*

In this case, there seems to be a mismatch between the intentions of Reidun's children and her actual use of the device. The motivation for the gift may well have been to enhance the grandmother's life, to allow her a freedom of movement, and to provide a link to others in the family. We can speculate, however, that the gift was also intended to facilitate caretaking on the part of Reidun's children and grandchildren.

Reidun had experimented with using the mobile telephone in order to increase her range of action. The experience of staying at the cabin was not what she had hoped for. After this, the mobile phone had seemingly fallen into disuse. Nonetheless, she also has the notion "if something should happen" near at hand. In this way, her understanding of the mobile telephone as a lifeline plays on some of the same themes as expressed by the other informants.

Another security-related dimension of the mobile telephone could be found in its symbolic use. Several women described using the telephone when they were alone in the city to indicate—sometimes falsely—that they were engaged in a telephone conversation. They reported using this strategy when they felt intimidated by the area in which they found themselves.

> **Grethe 19:** *I was alone in town once and there was a car that followed me and suddenly it screeched to a halt. And so I immediately took out my mobile and called my boyfriend, and they [in the car] left immediately when they saw the mobile.*
>
> **Moderator:** *So you see the mobile as a type of security?*
>
> **Grethe:** *Yeah, then I was glad that I had the phone.*

Other informants described the same technique. In one case, however, one informant described simply going through the gestures of calling.

> **Hanna:** *When I am alone, often I just call someone just to talk, if you know what I mean. If I am going to go home and it is late at night, then I often call someone to talk to them while I walk.*
>
> **Anniken:** *If I am alone in town, for example, I get out my phone and just start talking.*
>
> **Hanna:** *I think more that if something happens to me, then I can call for help, because if you get beat up or something like that and there is nobody to help you.*

Anniken's comments point to the use of the mobile telephone as a prop or symbolic icon for connectedness. Her comments indicate that the posture of speaking on the phone can be seen as part of a broadly understood gestural syntax. Anniken describes how she drew on this syntax—in a bogus form—in order to communicate to a potential harasser that she was in contact with others. Her comment shows how the mobile phone can be employed both as a communication channel and as a symbol of being connected. In turn, this provided a vicarious protection. This technique was useful for the women when they were alone at night in the city. In that case the mobile telephone is not even necessarily used in its function as an electronic terminal, but it is being used to communicate the idea to potential perpetrators that the woman is in touch with others and thus is not "alone."

The various techniques described here extend the perceived boundary of safety; that is, it provides a sense of security. The technology extends the range of activity into areas where "something could happen" or there may be an "emergency." This is, however, not complete freedom. While some of the notions of security were fanciful—such as Karl's idea of calling a helicopter while skiing in the forest—the informants often understood that limits on their freedom were determined by their distance from the local antenna, by the life of the batteries, etc. Informants displayed a nuanced understanding of the various systems and the coverage for the competing operators. The knowledge regarding coverage is, in some cases, quite detailed. It is mapped onto specific portions of the countryside and even down to parts of a cabin:

> *I have a cabin up near Nesbyen, up in the mountains there, and I had to move it up to the mantle. There it was good, but in the kitchen it didn't work.*

The failure of batteries or the poor coverage of certain areas means that the boundary is not absolute; rather, it moves and is, to a degree, unstable. It is not something on which we can rely absolutely.

The Mobile Telephone in Extraordinary Situations

The discussion up to this point has focused on what we might call mundane security or the use of the device in everyday life for situations that concern a limited number of people. However, mobile communication has also developed a role in broader catastrophes, notably the events of September 11, 2001, in the United States. There is a small literature on the use of mobile telephony during those events (Palen 2002; Rice and Katz 2003; Dutton 2003).

This literature discusses the way in which news regarding the airplane hijacking and the developing events was communicated via the mobile telephone. In the case of Flight 93, the plane that was recaptured by the passengers, the multiplicity of communication channels played a role in allowing the passengers to notify others as to the situation, gather news with regard to the breaking events, and validate information. Based on this they were able to organize themselves and overtake the hijackers.

Mobile communication devices were also employed during the events at the World Trade Center. In some cases, they allowed persons to send and receive information. However, these events—as well as the blackout in 2003—in some cases underscored the vulnerability of the radio-based communication systems. Loss of central transmission hubs and the loss of power to the system hobbles its ability to cope. This said, some of the rescue efforts at the World Trade Center were coordinated via mobile phones and wireless text-based devices.

It seems, however, that other, expressive functions were also central here. Mobile communication devices were used by those onboard the planes and in the upper floors of the World Trade Center to communicate with loved ones in attempts to comfort one another in the face of an unthinkable situation.

The emphasis in the calls was not on instrumental communication, but rather on the need to impart emotional support.[14]

Those who were unsure as to the fate of their loved ones also used the mobile telephone. Rice and Katz report on the use of the mobile phone as a way for family members to contact each other and check on their status (2003). Here we see parallels to the situation in Israel, where the mobile telephone is used to check on others' status in the wake of attacks (Lemish and Cohen 2003). Another important function of the mobile communication devices was that those caught in the upper floors of the office towers and those on the planes used them to say final goodbyes to loved ones. Thus, the events of September 11th as well as the use of the device in Israel underscore its role as a type of lifeline and as a link between loved ones. In addition, the representation of these uses in the media adds a level of legitimation to the lore surrounding mobile communication.

Many of the citations reported earlier were in interviews from before the September 11th attacks. Thus, the mobile telephone was broadly seen as a contribution to security even before the events of the 11th of September. It was seen as a useful way of being "on call" or for summoning help in the case of those with chronic problems. In addition, it was seen as a way to help us through what we might call mundane crises. The graphic and heart-rending stories of victims using the mobile telephone in a widescale disaster add a further dimension to this well-established perspective.

This gloss is played out in two ways. First, mobile communication has been, and will continue to be, used to assist people in various types of real emergencies. It will allow those with heart conditions to operate in a wider sphere, it will be used to summon assistance when our cars break down, and it will be used to provide us with a way of indicating our connectedness when we feel intimidated by the situation in which we find ourselves.

At the same time, its use can, and probably will, be overplayed. The reliance on mobile communication cannot provide absolute security. The precision of enhanced emergency telephone services can be misleading. The reliance on a mobile phone to intimidate attackers can be a false security. There are areas of poor coverage where, should you have a medical emergency, you cannot reach the outside world. We see here the interaction of the technical capacities of mobile communication and their application in everyday life. It is exactly in this type of situation that folk knowledge is drawn upon, embroidered, and legitimized or negated in order to determine where the boundary between safety and foolishness lies. Thus, we can expect to see the codification of folk understandings as to where, when, and to what degree we can rely on mobile communication in emergencies.

This discussion will also take place on a moral plain. On the one hand, there is a discussion surrounding the desirability—or even a type of moral responsibility—of having a mobile communication device when we are out. Indeed this has already been formulated in the oft-cited phrase "in case anything should happen." We see this in the comments of the fellow who was pinned down by the irresponsible target shooter. He continued his comments by saying:

> *This emergency sold me on the value of a cell phone as a security item. I carry mine even if I do not want to take calls and don't expect to make them, as it could save someone's life. I would rather my wife carry a cell phone than a gun.*

At the same time, whenever we rely too heavily on the mobile telephone to no avail, there will be the opposite reaction. In those cases, that is, in situations such as when the hikers were stuck in a snowstorm and their mobile phone could not reach a base station, or you are marooned with car problems and have an uncharged battery in your mobile phone, the victims will likely be the subject of moral statements that they relied too heavily on what, after all, is an unstable technology.

In sum, safety and security have become a well-engrained part of our social image of the mobile telephone. This image is a part of the way in which we have legitimized the device. It is used in our understanding of how we integrate, or perhaps domesticate, the mobile telephone into everyday life situations. These legitimations can, for example, be employed in discussions regarding its purchase or use or to counter other, more critical, portions of the lore.

Further, the material here describes something of a paradigm shift. We can trace a shift in the social understanding of mobile communication. The technology has gone from being a symbolic prop for the rich to having a central, albeit often passive, role in the lives of many people. It is important to understand that the legitimations are social constructions. They are based on concrete situations such as the time your friend was stranded in a potentially dangerous situation and the mobile telephone came to the rescue. These events have been reformed into a broader understanding of the mobile telephone as a type of umbilical cord.

Diminution of Safety

While there is the sense that the mobile telephone provides security and safety, the picture would be incomplete if I were not to discuss the ways in which mobile communication can potentially *reduce* our safety. It is clear that the mobile telephone can be used to organize illegal activities of criminal gangs (Godø 2001; Lien and Haaland 1998).[15] Indeed the terrorists who organized the September 11th attacks through a thinly connected network (Krebs 2001) relied to some degree on mobile communication.[16] In this case, mobile telephone records show that the device was used by the terrorists on the day of the attack, possibly for coordination (Dutton 2003; Rice and Katz 2003).

Driving and Mobile Telephone Use

Moving from the world of international terrorists and underworld figures, we see there is also a significant body of literature around the issue of mobile telephone use while driving. This includes experimental studies as well as epidemiological analyses of accidents (Cain and Burris 1999).

The experimental studies generally rely on simulated driving situations in laboratories. They have examined the effects of auditory distractions (ICBC 2001), talking on the mobile telephone (Strayer and Johnston 2001), visual capacity (Strayer *et al.* 2003), and various combinations of these elements (Recarte and Nunes 2000; Burns *et al.* 2002). The general finding here is that the mobile telephone increases reaction time and that it demands attention that better could be afforded driving.

The epidemiological studies examine the use of mobile telephones in actual traffic accidents. Perhaps the most influential of these studies was carried out by Redelmeier and Tibshirani (1997). The authors gathered data from 742 persons who were involved in auto accidents where there was no personal injury during a period in 1994–95. Telephone records were collected and examined for telephone activity in the minutes before and after the reported accident. The results show that 24% of the participants used the telephone in the 10 minutes preceding the accident. Further, they found that use of a "hands-free" device had no effect on the potential for being involved in an accident. In addition, younger drivers were more prone to "mobile phone" accidents than older drivers, and accidents were more likely as traffic speed increased. As a type of fig leaf, the material shows that almost 40% of the participants called to emergency services via the mobile telephone after their accident. Similar epidemiological analyses have been carried out for those involved in fatal accidents in Oklahoma and in a broader study from New York (Cain and Burris 1999).

While the specifics of the studies are often lost, the general picture is not lost on those who participate in interviews.

Georg: *There was one that drove into me from behind and was calling on the mobile. It didn't hurt the car, so it wasn't a big deal, but ...*

Harald: *There is going to be a law ...*

Karl: *That is right.*

Bjørn: *But now they are giving out some statistics about accidents caused by mobile telephones, at any rate for people that admit it.*

Karl: *I think that there are many unreported accidents.*

Bjørn: *I think so too.*

The sense here is that the mobile telephone is a background element in many traffic accidents. Others have the general notion that mobile telephones are dangerous but that their effects can be modified via the use of "hands-free" devices.

Oda 18: *There is maybe a difference between having one in the car and talking while you drive. At any rate when I was out driving and have seen people talking [on the phone], they swerve back and forth. You don't drive behind them, you drive another way. If you have hands free, then it is ok, you know. But if you need to change gears and things, then you can't do it with a mobile telephone without letting go of the wheel. That is not too safe.*

Yet for others, their perspective balances the potential danger posed by the mobile telephone with the notion that we can also use it to call for aid in emergencies.

Arne 17: *They have done studies about that, in relation to when people use the mobile phone when they were talking when the accident happened. There is a large majority that thinks that it is safer to have a mobile in the car than not, since people can call if there is an accident if something has happened.*

Turning to quantitative data, the sense that using the mobile telephone while driving is dangerous is particularly strong in Denmark, the Netherlands, and the United Kingdom. Using the same Europeanwide data set described earlier, we find that more than 70% of the respondents in those three countries completely agreed with the statement "A mobile telephone is a hazard when one drives." By contrast, there was relatively weak support for the statement in Germany and Norway, where less than 50% of the respondents agreed completely. We can perhaps clarify the low support in Germany by pointing to the special automobile/autobahn culture there, and the Norwegian case may be explained by the need for access to others, given the sparse population of the country.

For the whole sample, women were significantly more likely to agree with the statement than men. The data shows that while 64% of the women were in complete agreement, only 59% of the men shared the same perspective.[17] Further, agreement with the statement increased with age. Only 57% of those under 25 agreed completely with the statement, as compared to more than 71% of those over 67.[18]

Although there is a strong sense that use of the mobile telephone while driving is dangerous, there is nonetheless a strong motivation to use it in just that situation. Cain and Burris report that about one-third of all mobile phone owners say that they use the device in the car on a somewhat regular basis (1999). Thus, the question is why mobile communication has become a part of driving.

In a nicely turned phrase, Fortunati notes that the mobile fits into the folds of everyday life. The thought is that we use the device in times and places that are otherwise unoccupied. We use it when waiting for buses, sitting on trains (often to the annoyance of our fellow travelers), etc. From this perspective, driving time is a type of gray zone in the minds of many people. It is often seen as free time when we are sheltered from other demands on our time. While driving requires attention and while unexpected situations arise, it is also extremely regulated and very routinized.

To put this into a sociological context, driving is perhaps that area of everyday activity which in we must internalize the broadest range of legally codified norms. Rules as to who proceeds first through left-hand turns, who must yield to whom, the speed at which we can drive, etc. are all proscribed. Beyond the formal laws, there are, of course, the informal norms and courtesies that we observe while driving. Drivers' exchange of nods so as to yield one another the right-of-way, the careful interpretation of pauses, and the use of turn indicators fall into this class of behaviors. Finally, there is a morally emphatic vocabulary that can be used to describe—and perhaps confront—drivers who break the rules by, for example, cutting off others, weaving in traffic, or snapping up parking places. This underscores the fact driving is an activity that is heavily governed by norms. From a sociological perspective, laws, rules, norms, and moral judgments can be seen as markers of activities in which overt social coordination is important. Indeed, in the case of driving it is essential to avoid accidents. At the same time, the extremely routinized and norm-governed nature of driving means that we are open for other types of engagement.

The automobile has been an area where alternative forms of diversion have been allowed.[19] Thus, it is not surprising that people often describe car time as quasi-open time, time in which they can attend to other activities. This comes through in the comments of Ida and Vera, who enthusiastically describe the ability to take care of small tasks via the mobile telephone while commuting.

Ida: *To do several things at once is satisfying. I got sooo much done! you know?*

Vera: *If you have a reminder list of 20 things and you can check off three of them on the way to work, oh that is so nice.*

Ida: *It is like fresh air!*

Interviewer: *What kinds of things?*

Ida: *Things at work where you have to call to plan meetings, a dentist appointment.*

Vera: *Things that you haven't tackled, things that you have been trying to do for the last three days. It is a, a relief [to get them done].*

The same sense comes through in the comments of Harald.

I use it [the mobile telephone] in the car, but I use it hands free. I have a thing in my ear and a microphone that hangs there. I can do many things in the car. We have a factory in Drammen, for example, so I drive from Oslo to Drammen quite often and other places also, and that is time that I can use more effectively and make appointments and things like that. And if I don't want to have it on, I turn it off. So it is very effective in the car, if you have hands free. (Harald)

Ida, Vera, and Harald are typical persons who have to manage their lives on several fronts. They have jobs that require attention to a broad range of tasks and details. In addition, they have active family lives that have another, overlapping set of demands. Thus, in Vera's words, the ability to "check off" a couple of items on the way to work is seen as a positive thing. From this perspective, the mobile telephone can make time in the car more effective. Obviously, this perspective is not without its problems.

Thus, there is a conflict. On the one hand, people are familiar with the evidence that driving and talking on the phone is a potentially dangerous combination. Indeed people can often cite the general contours of this research. In addition, many people had directly experienced accidents involving driving and mobile telephones, or at least they knew of people who had had such experiences. Nonetheless, the quasi-open nature of driving a car means that we are motivated to use the time for various organizational or social interaction. Based on this, one suspects in spite of the evidence that mobile telephony and driving do not mix, this phenomenon is likely to become a fixture in our everyday lives.

Personal Privacy

Another area in which there is potentially a diminution of safety when considering the use of the mobile telephone is that of personal privacy (Green and Smith 2002). At one level the mobile telephone is just another element in what Gary Marx has called the "surveillance society" (Marx 1988). According to Marx there is a growing ability to trace and cross-reference our activities via our various digitally assisted transactions. The resulting picture might provide insight into our medical condition, buying habits, or particular demographic situation.

This scenario generally examines how large, "big brother"-like entities such as governments or large corporations might use the assembled information vis-à-vis the individual. Another issue, however, is the enhanced ability to follow the movements and situations of specific individuals. Various location-transmission devices—some are integrated into mobile telephone system, while others are standalone devices—allow us to follow the location and movements of others. Indeed, at the insistence of insurance companies, location-transmission devices are sometimes buried in the bowels of expensive automobiles in case they are stolen.

Less expensive and easily concealable location-transmission devices allow us to follow the movements and location of pets, children, or, more ominously, for example, former boy-/girlfriends. On the positive side, location-transmission devices can be seen as contributing to the general safety of society, in that they allow quicker response times in case of emergencies. At the same time there is a Faustian dilemma, because they open the possibility for various types of lurking and stalking, an issue that often affects women more than men (Tjaden and Thoennes 1998). Yet another issue here is the use of mobile telephone–based photography. While the average urbanite appears on hundreds of video screens daily, the introduction of mobile telephone–based photography whereby one can easily transmit the images to the Internet, and indeed the world, adds a new dimension.

Conclusion

In summary, the mobile telephone has found a niche as a device that provides safety and a sense of security in a variety of situations. For those who have chronic conditions, the device allows a broader freedom of movement and a sphere of action. In addition, the device is seen as being useful in acute problems, ranging from punctured tires to life-threatening situations. This aspect of mobile telephony has been abstracted to the point that the mobile telephone symbolizes security. Not only do we feel secure when we have the device in our pocket, but the material shows that the very form of using the mobile phone is seen as a way of defending against potential assault.

That is the positive side of the coin. They can also threaten public safety; that is, the mobile telephone can be used for various criminal activities. In addition, using the phone while driving has been shown to increase the possibility of having an accident. While the link between driving and telephoning is commonly understood, it is also commonly ignored. The use of the mobile telephone has found a place in the free spaces of our schedules. Using the phone in these situations allows us to attend to the small coordination chores we would otherwise do in more stationary settings. Driving the car is such a space, and the urge to take care of a few errands is often weighed against the relatively low level of attention needed to drive a car. In this way, we can account for the use of the device in this situation. As we will see in the next chapter, mobile telephony is an ideal technology for coordinating everyday affairs. It allows for dynamic planning, thus making group interactions more robust and independent of place. However, in the context of safety and security, this characteristic can bounce in the wrong direction.

We can place these findings into a broader discussion of the social consequences of mobile telephony: Safety and security have become a well-engrained part of technology in the social consciousness. The technology has changed the way we think about dealing with various types of risk and mediated their effects in certain ways. It has afforded us a kind of freedom and a sphere of action that is much larger than that of the landline telephony system. In addition, the technology has introduced us to new forms of risk. In the elaboration of this dialectic, we see an ongoing dialogue in the definition of the mobile telephone here. At one turn of the discussion, the mobile telephone is there "just in case something happens." At another turn, it is the cause of a fender-bender—that is, perhaps reported via the same mobile telephone that originally caused the accident. At yet another time, the mobile telephone allows us to react quickly to an unexpected situation, and at the next turn of the cycle it is used by gangs to organize the distribution of heroin or by someone to stalk a former lover.

Generally, people have a clearer focus on the more positive dimensions. The mobile telephone has been understood, at least partially, in these terms. Our adoption, our use, and to some degree our sense of self and the way others see us revolve around items such as this. For some people, to play on Goffman's phrase, the device becomes a part of our "possessional territory." That is, the mobile phone is an object that can be identified with the self and that, in turn, helps others to identity and perhaps characterize the individual (Goffman 1971). At a broader level, the general recognition of the mobile telephone as an aid in protecting our safety points to the broader legitimization of the device and its integration into the social order.

4

CHAPTER

The Coordination of Everyday Life

Introduction

Einstein is reported to have said: "The only reason for time is so that everything doesn't happen at once." While he was probably thinking of lofty things such as relativity, the speed of light, and other celestial mechanics, there is also a practical, everyday side to his comment. Back here on earth, time and timekeeping are essential elements in the way we plan, arrange, and coordinate our affairs. In the words of Louis Mumford, "The clock is not merely a means of keeping track of hours, but of synchronizing actions" (Mumford 1963, p. 14). Time as mediated via timekeeping devices tells parents when they need to deliver and pick up children at day care, rush to work, turn up at meetings, buy tickets for films, rearrange activities because of traffic jams, get their kids to piano lessons, agree on who will pick up the car at the garage, and innumerable other interactions that make up everyday life. We are constantly saying things like "Can we meet in an hour to talk about your proposal?" "The concert starts on Thursday at eight," or "I'll be there in 20 minutes." All of this is done with the taken-for-granted assistance of time and timekeeping. Time and mechanical timekeeping represent perhaps the most well-entrenched "technology" for the coordination of social interaction.

This social coordination, mediated through our sense of punctuality, allows for more or less harmonious interaction. From a social perspective, time and timekeeping are generalized, robust, and easily accessible to all. Beyond being a functional system for the coordination of interaction, we have also developed rituals, manners, and courtesies associated with timekeeping. Thus, there is also a moral dimension to observing time, to being late, or to being obsessed with time.

While time and timekeeping are well-established methods for coordinating social interaction, I wish to assert that we have also started to use the mobile telephone in this context. That is, the mobile telephone has started to change the ways in which we organize and coordinate our everyday lives.

As the mobile telephone becomes ubiquitous, it competes with and it supplements time-based social coordination. In essence, we begin to move away from the parallel interpretation of a common metering system, i.e., time, and replace that with the possibility for direct contact between those who are coordinating their interactions. Instead of relying on a mediating system, mobile telephony allows for direct contact that is in many cases more interactive and more flexible than time-based coordination. I quickly want to add a caveat: The mobile telephone will not replace timekeeping in all situations. However, in the case of small-group interactions, the mobile telephone has made serious inroads into the hegemony of time-based coordination.

Mobile telephony has introduced the ability to call quickly or to "text" to others and change plans when new exigencies arise, using a type of microcoordination, or the nuanced management of social interaction via the use of mobile communication (Ling and Yttri 2002). Thus, at a very basic functional level, mobile telephony allows for a tighter microcoordination of our social interaction.

Many studies have shown that traditional landline telephony is used primarily to carry out instrumental activities, such as coordination. In a macroanalysis of studies done in France, Germany, Japan, Korea, and the United States, LaRose found that between two-thirds and three-quarters of all calls have to do with instrumental activities, such as confirming appointments and organizing affairs (LaRose 1998).

The mobile telephone extends this possibility. It allows us to plan and replan activities — any time and anywhere — to a greater degree than with the traditional landline telephone. Unlike previous communications systems, where we were bound to a certain location, use of the mobile telephone means that we need not be at a specific station, node, or geographical location to receive information. Perhaps more importantly, we do not need to know the location of our interlocutor. This can increase the efficiency of planning and make it easier to carry out everyday tasks (Lange 1993; Cooper *et al.* 2002). Indeed, one can assert that this is the greatest social consequence of mobile telephony.

The coordination of everyday life via the mobile telephone is popularly seen as one of its main advantages. Data from a Pan-European study shows that individuals see the usefulness of mobile telephony in the coordination of everyday life.[1] In general, respondents were quite positive about the use of the mobile telephone for planning and social coordination. The data show that 69% of the respondents thought that the mobile telephone was somewhat or very helpful to coordinate family and social activities (Figure 4.1), and 70% agreed with the idea that the mobile telephone allows them to stay in steady contact with family and friends (Figure 4.2). By way of comparison, somewhat fewer respondents (64.5%) thought that the mobile telephone was a disturbing influence, and quite a few more (96.1%) thought that it was useful in an emergency. However, the data show differences between groups. In general, we find that younger persons,[2] and to a lesser degree men,[3] had a significantly more positive attitude toward use of the mobile telephone to "coordinate family and social activities." The same is generally true of the statement describing the use of the mobile telephone to "stay in steady contact with family and friends." Younger persons[4] and men[5] were significantly more likely to agree with this statement.

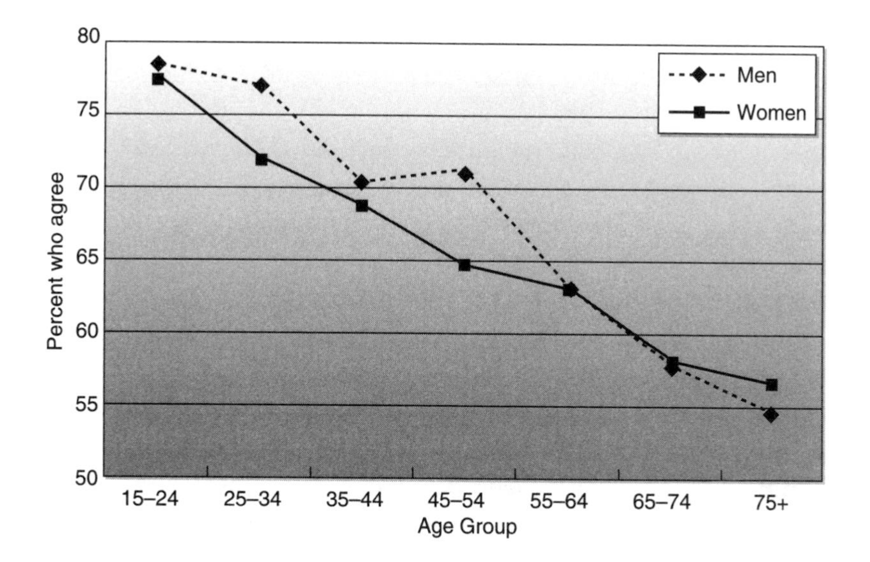

FIGURE 4.1 Percent of persons who agree with the statement "The mobile phone helps one to coordinate family and social activities." From the Pan-European EURESCOM P903 study of mobile telephony and Internet use, 2000. (Note truncated scale.)

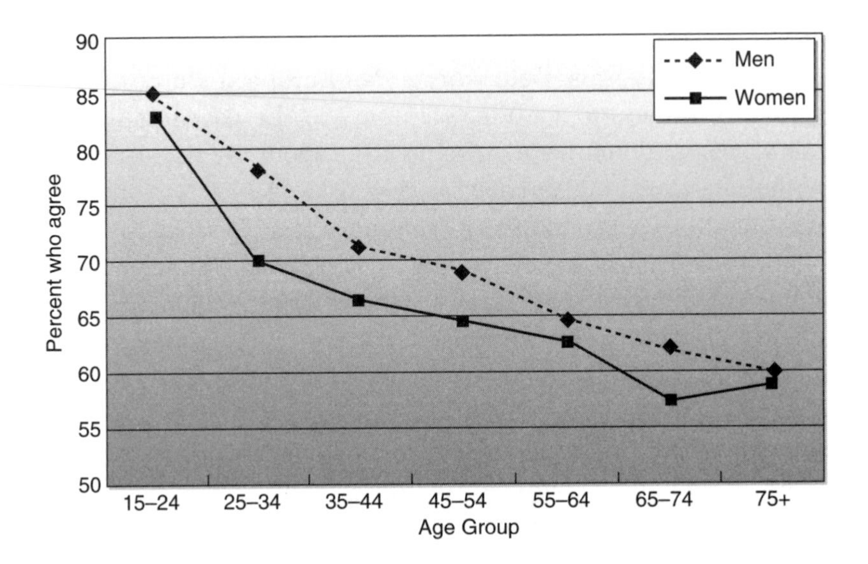

FIGURE 4.2 Percent of persons who agree with the statement "A mobile telephone helps one to stay in steady contact with family and friends." From the Pan-European EURESCOM P903 study of mobile telephony and Internet use, 2000. (Note truncated scale.)

Thus, while it is often parents, and more specifically mothers, who can benefit from microcoordination, the groups who are most positive to it are younger persons and men. This may mean that this capacity will be fully developed only as the current generation of young adults moves into the position of being parents faced with the need to be seemingly everywhere at once.

In this chapter, I will look at social coordination with the use of timekeeping devices and mobile telephony. While these two technologies may seem to be quite different, they have their similarities. Timekeeping for coordination — in the form of watches — is indeed a personalized technology that has found a stable and taken-for-granted place on the body.

The mobile telephone can be seen in the same light. It is a personal coordination technology, just like the watch; however, it is not yet taken for granted, nor has it found its final locus on the body. I will examine the various dimensions of social coordination and how mobile telephony is in some ways challenging and in some ways complementing time and timekeeping as the medium through which we coordinate our everyday affairs.

Social Coordination

Coordination is a common everyday activity. As already noted, we are always agreeing to meet, planning vacations, organizing visits, managing schedules, synchronizing activities, arranging affairs, and orchestrating events. The need to coordinate activities is a basic social function, particularly in contemporary society, characterized as it is by distributed housing, automobile-based transportation, and complex activity patterns (Townsend 2000).[6] This is particularly true in modern motorized urban settings, where activities and functions are often highly specific and distributed. To take this out of the realm of sociology-speak, a characteristic of contemporary society is that we often have a variety of tasks and appointments spread around a city. One child may be at day care, a second several miles away in an elementary school. Shopping is off in one direction, your job is in a second, and your dentist appointment is in yet a third.[7] Thus, there is an almost inherent need in modern society to coordinate our activities and movements.

If we decompose the concept of coordination, we lay bare several important dimensions: (1) the number of people involved in the activity, (2) the period between planning a meeting or an event and its execution, and (3) the degree to which the planned activities are simultaneous and collocated. Obviously there are also power issues associated with coordination. Some people are given the prerogative to include or exclude others in activities or to determine their tempo, sequencing, and placement. There are several tools at our disposal when coordinating activities. Finally, coordination of activities can make obvious the power relationship between individuals.

Social coordination can involve different numbers of people. We can coordinate activities with one other person (a tête-à-tête or meeting an old chum for a beer), with a small group (a cell of conspirators or soccer practice for our child), or with thousands (such as a political campaign rally or the invasion of Normandy on D-day). Further, there can be different delays between the planning of an activity and the activity itself. We can coordinate activities for the immediate future ("Ling, come to my office, ASAP!"), for the more distant future ("Are you up for tennis next week?"), or for years hence ("When you retire, please consider moving to the Rockies"). Finally, the coordinating parties can plan events that are both simultaneous and collocated (e.g., several friends meeting for a cup of coffee) or activities that are consecutive and dispersed ("First I will get the children at school and then take the train to my parent's house. You will be staying at home. So remember, in the evening you have to water the plants and then call us after we have gotten to my parent's house.").

There can be power differences between the parties who are coordinating activities such that one partner has the prerogative to demand things of another. The coordination can also employ various tools and technologies. These can include simple rules ("If it snows we will meet at my house"), various forms of mediated interaction (such as the use of the mobile telephone or e-mail), and usually, though not necessarily, the reference to time in one or another form.

While the canvas upon which coordination can take place is quite large, it is perhaps useful in the context of this chapter to draw it in a bit to the scale of small-group interaction in the near and middle future. After all, it is not often that we plan a landing at Omaha beach — though some family vacations might lead us to believe the opposite. In everyday life, we often need to plan interaction with a limited number of others in the relatively near future. While this type of planning or coordination may be part of a larger effort, the encapsulated event has its own dynamics and identity. It is in this context that time is an essential tool, and it is into this context that the mobile telephone and microcoordination have found a niche.

The demands for rapid and geographically dispersed coordination of small groups became more acute due to the rise of transportation systems and the differentiation of social functions (Beniger 1986). The development of transportation systems has perhaps been the single greatest factor effecting the development of urban areas. Starting with streetcars and trains and followed by the development of the automobile, cities have changed from being a core city surrounded by a small ring of dwellings to being a vast suburban system of highways, shopping areas, and strip developments. Since the mid-1900s the automobile and its range of action have had one of the greatest effects on the development of urban areas (Crawford 1994; Grahm and Marvin 1996; Dyckman 1973; Hall 1973; O'Connor and Maher 1982; Thorns 1972).

In this context, the coordination of our activities and movements with others has taken on increased importance, a type of critical imperative. Effective coordination minimizes uncertainty regarding interdependent activity, when and where common resources will be accessed or consumed, and when transition between activities will take place (Blount and Janicik 2001). As noted earlier, both time-keeping and telecommunication have been used in this capacity.

It is also important to note that there is a gendered dimension to social coordination. Vestby found that working parents, and in particular working mothers, exert a type of loose control over their children's home-based after-school activities. There is recognition that this control is limited, but there is at least the geographic certainty that the child is at home (Vestby 1996, p. 75).[8] The adoption of mobile telephones by parents in some ways extends the ways they interact with their children. Rakow and Navarro have described the effect of mobile telephones as "remote mothering" (Rakow and Navarro 1993). It has been noted, for example, that men use the telephone to make specific arrangements, while women use it to integrate a social network.

The telephone is a technology that makes women accessible as caregivers and thus can enforce and extend the private sphere. Claisse and Rowe report that women are often the "social administrator" for the home via their role as the one who has the responsibility for answering the telephone. They indicate that the telephone was used to manage, inform, coordinate, and discuss, and that women had a central role in this (Claisse and Rowe 1987, pp. 213–214; Lange 1993; Bakke 1996).

Mechanical Timekeeping and Social Coordination

One of the most central contributions of mobile telephony is its potential to allow for more nuanced coordination of everyday life. To better appreciate this, it is important to examine the role of time and timekeeping in social coordination so as to understand the way in which mobile telephony has played into this aspect of everyday life.

Solar-based timekeeping has long been a social fixture. Chaco Canyon's Casa Rincada in the United States, the temple of Kukulcan at Chichén Itzá in Mexico, and Stonehenge in the United Kingdom point to this (Eriksen 1999; Aveni 2000; Zerubavel 1985). These structures were, however, used for marking the solar cycle for ceremonial purposes linked to agriculture. They were not associated with the humdrum coordination of everyday life in the sense described here.

There were two developments needed before timekeeping could be used for everyday social coordination at the small-group level:

1. Most importantly, broad access to robust devices that allowed reasonably precise time keeping

2. An agreement on the basis for the synchronization of these devices

Interestingly, it was not until the late 1800s that these two conditions were achieved.

The Development of Mechanical Timekeeping

Starting in the 13th century we see the development of reliable mechanical timekeeping (Landes 1983; Zerubavel 1985). The earliest record of a weight-driven clock is from 1283 (Andrewes 2002). Landes has suggested that one of the specific motivations for the development of what we know as a clock was as an aid in maintaining the cycle of prayer in medieval Benedictine monasteries (1983).

These monastic devices were adapted to "civil" purposes during the 1300s, when it seems there was a veritable boom in the instillation of stationary tower clocks. Any self-respecting town had one. The construction of a clock tower was a type of public works initiative that, in addition to coordinating municipal institutions, provided an outward sign of the locality's wealth, innovation, and the strength of its administrative bodies. The town clocks could include calendars and faces showing the alignment of the heavens and even play out various types of mechanized theater.[9]

The introduction of mechanical timekeeping during the later medieval period started the regulation and coordination of all forms of social interaction. As stated — somewhat critically — by Mumford, "Mechanical periodicity took the place of organic and functional periodicity" (1963, p. 198). Town clocks were used to regulate market time, the start and close of town meetings and schools, the length of daily pauses, the time allowed for various types of sporting events, the length of various punishments and tortures,[10] and, of course, the cycles of religious activity (Dohrn-van Rossen 1992). The limitations, however, are also obvious. Town clocks were rather hefty centralized instruments full of gears and weights and were ultimately designed to ring bells.[11] They were by and large designed to coordinate the activities of larger social institutions — church services, town meetings, and the like. Further, this form of coordination was useful only for those who were within the audible or – eventually – visual range of the clock itself.

Moving from stationary to portable timekeeping was the next technical development. The need for accurate portable timekeeping was acute, particularly in the area of marine navigation.[12] The development of the pendulum allowed the use of smaller devices that were portable, in that one could move them from place to place, but they were useless in transit and needed to be resynchronized after each move (Andrewes 2002). John Harrison, a local clock maker without formal education, was able to develop the first truly portable and accurate timepiece in the mid-1700s. His marine chronometer was accurate enough to allow the precise navigation of a ship, something that was not possible to that point (Sobel 1996; Browowski 1973).

With the development of the robust, mobile, and accurate marine chronograph, mechanical timekeeping began to address the first of the two prerequisites noted earlier, i.e. the development of a robust device that allowed reasonably precise timekeeping. Nonetheless, this did not mean that timekeeping was generally available. Pocket watches were extremely expensive, handcrafted devices. Indeed, the development of Harrison's final chronometer took decades. Even the fitting of the works of a common pocket watch required careful skill. You needed individually to file, adjust, and readjust the various gears and springs for each timepiece. Because of this, the clock industry was largely resistant to mass production long after other types of machinery were mass-produced.

It was not until the 1850s that the Waltham Watch Company succeeded in mass-producing a pocket watch, thus reducing the cost of production. As a coincidence, the American Civil War induced the general adoption of the watch. The device provided its bearer with a claim of social prestige (Levine 1998; Andrewes 2002). An added bonus was that pocket watches could be used to coordinate military maneuvers (Dohrn-van Rossen 1992; see also Kahlert *et al.* 1986).[13]

Thus, from the 1850s you could regularly find portable timekeeping devices that were relatively robust, inexpensive, and precise. Given these developments, it was generally possible to use time as a coordination technology, particularly in coordinating local activities. You could agree to meet with a lover or to initiate a cavalry charge at a particular moment in time with the knowledge that the participants' respective timepieces would instruct the bearers as to a simultaneous point in the future when the tryst or the bloodshed would take place. That is, they were coordinated. However, if you were to coordinate activities across broad distances, such as is needed when running a railroad, another issue arose: the standardization of time.

The Standardization of Time

In the 1850s, people synchronized their watches by using the solar noon in their specific location (Levine 1998). However, the use of solar time, even if it was marked mechanically, meant that it was not possible effectively to coordinate systems that spanned larger geographical areas. With the development of railways, the traditional system of using local solar time as a standard became problematic, particularly in countries such as the United States that have an east–west orientation.

Before the standardization of time zones, each railroad company used the local time of its headquarters to synchronize its trains. If a company based in Philadelphia and another based in Columbus each had a train scheduled to leave Pittsburgh at noon, the Philadelphia-based train would actually leave Pittsburgh 32 minutes before the Columbus-based train. This is because solar noon in Columbus is approximately 32 minutes after that in Philadelphia. Thus, a clock alone, regardless of how precise it was, would not help the poor traveler to understand when he or she had to be at the train station. In the mid-1880s, there were as many as 50 competing times standards used by various railroad companies in the United States (Blaise 2000).

Beyond being an annoyance for travelers, the system of using local standard times was a threat to safety, since it was extremely difficult to know when two trains were using the same tracks. During this period there was a series of train wrecks that, more than anything else, spurred the development of more generalized synchronization efforts (Beniger 1986, p. 228; Blaise 2000).

All of this pointed to the need for replacing local solar-based time with standardized time zones. This was achieved in the 1884 at the so-called "Prime Meridian Conference" in Washington, D.C. It was here that Greenwich, England, was established as the location of the prime meridian. In addition, the globe was divided into 24 zones of roughly 15° each. There was lively debate as to where the meridian should be placed. Rome, the Great Pyramid, Paris, Berlin, the Vatican, Cadiz, Jakarta, and a host of other locations vied for the honor of "hosting" the global point of reference. However, given the fact that much of the marine navigation literature already had adopted Greenwich and that the opposite meridian, i.e., the International Date Line, conveniently passed through much of the Pacific Ocean, it was decided to use the "meridian passing through the center of the transit instrument of the Observatory of Greenwich as the initial meridian for longitude" and further that "from this meridian, longitude shall be counted in two directions up to 180 degrees, east longitude being plus and west longitude minus." After wounded national prides had time to heal — as well as those local sensibilities who mourned the loss of their own "local time" — this became the accepted standard (Palen 1998; Levine 1998; Zerubavel 1985; Blaise 2000).

Only at this point can it be claimed that there was general access to robust and reasonably precise timekeeping devices and broad agreement as to the basis for synchronization. Thus, time became the de facto, and indeed taken-for-granted, mechanism with which to coordinate various forms of social interaction.

Time has become the meter against which we control the ebb and flow of our lives. It is reasonable to assume we are "in sync" with others. We can assume that others generally know that to meet in a "half an hour" is a meaningful utterance that will result in a common goal. We can assume that others interpret the phrase "We can meet at the bar on 3rd Street after work at about 5:30" to have the same meaning for the partners in the conversation and will result in drinking a beer with a friend at about 5:35 — provided they don't call on their mobile phone to say they are delayed. A person in Oslo and another in Denver can agree to have a telephone meeting at 7:00 PM Oslo time. Further, the presence of clocks, both in public places and those we wear, mean that we must truly be isolated if we do not want to know the time of day.

The development of inexpensive quartz clocks has meant that timekeeping has become even more accessible. Extremely precise clocks have been included in everything from microwave ovens to video players. They are so widespread that even if your own watch does not function — or in the case of a video player if you do not understand how to program it — there are myriad alternatives freely available. Indeed we can speak of a tyranny of time (Eriksen 2001; Bell 1980; Mumford 1963) and different psychological (Zerubavel 1985) and social (Hall 1989; Aveni 2000) understandings of when we are on time and when we are late.

The Etiquette of Time and Timekeeping

Beyond being a method with which to synchronize interaction, respect for time, promptness, and tardiness have been integrated into our sense of manners. Etiquette manuals describe the problem of arriving either too early or too late for a social engagement (Bech 2002, p. 52). The need to respect time is stressed in etiquette guides for children (Young and Buchwald 1965), onboard ships (McDonald 2002), at concerts, and at dinner parties (Post 2002). With reference to being late, Lienhard (2002) says, "At 10 minutes I owe you an explanation: 'The freeway exit was closed. I had to go four miles out of my way.' After 20 minutes I have to make a full and serious apology. After 40 minutes I'd better not show up at all." The sense of punctuality, of course, varies between and even within cultures (Levine 1998; Hall 1989) and organizations (Blount and Janicik 2001).

This points to the degree to which time and timekeeping are a part not only of the instrumental integration of society but also of the way in which we steer through the moral shoals of life. In this sense, time and the social knowledge of time have become entrenched and are now treated as social fact, in the Durkheimian sense. Its importance is seen in the courtesies and manners associated with time. It can be seen, for example, in how early or late we can be for an appointment.

It can be seen in the way that we ask forgiveness when arriving late for an engagement. It is also seen in the way that we adjust our schedules to better fit the agenda of those who are in relatively powerful positions.

Thus, beyond its use in the synchronization of the activities of larger social institutions, it is also a morally infused dimension of small-group interaction. Indeed, learning the tempo of various events is a part of the acculturation process.[14] In the United States for example, if you are invited to a party at seven in the evening, you show up at 7:15 or 7:30. To arrive too early — that is, to arrive on time — is interpreted to mean that you are slightly too eager. Often we are instructed to arrive "sevenish," the implication being that to arrive precisely at seven means that we will meet a partially composed host and the other wallflowers who have nothing better to do than to show up at parties on time. In Norway, by contrast, you strive to arrive precisely at 19:00 so that you have time to take off your jacket or coat and to array yourself before the actual start of the party. To do otherwise is to be inconsiderate to your host and the other partygoers. The thinking is that everybody needs to be there before things can get started and the soufflé can be served.

These issues point to the fact that time and timekeeping are morally imbued. It is not simply a cold, rational system of coordination; it is also used in our estimation of others. It is included in the way we judge the social competence of those with whom we interact. In this way, time and punctuality are an aspect of etiquette. They are part of the informally codified norms associated with manners and courtesy. Etiquette can be described as calibrating our behavior based on how we think the other (either individually or collectively) would want to be treated (H. D. Duncan 1970, p. 266; see also: Geertz 1972, p. 290; Gullestad 1992, p. 165).

While the exact way of affording courtesy may differ from person to person and from culture to culture, a general form is that we are showing respect for the other and the other's station. While at the outset this might sound like the Golden Rule, it goes beyond this, in that our sense of power comes into play. Thus, we treat a king or a president as we think a king or a president would want to be treated. Indeed, the more powerful we are, the more we have layers of courtesies available with which to protect ourselves. In addition, those in powerful positions can choose to extend or to deny others courtesies. The boss can choose not to return your call, but you cannot usually choose to neglect calling the boss back. The president can keep others waiting, but others cannot keep the president waiting.

We have a repertoire of devices and techniques that can be used to effect courtesy. All cultures have quasi-formalized versions of manners, a codex of rules we are taught to observe. These can include, for example, which fork to use, what to do with a finger bowl, how punctual we must be, and how we manage the situation when we are late. It is here that we find a whole industry of authors who collect the various rules, protocols, and procedures into books describing courtesy. Thus, we can refer to Amy Vanderbilt, Judith Martin, or Toppen Bech for their suggestions as to the seating arrangements for a dinner with a bishop, or what to wear when invited to a reception on a yacht or at a day care center.

Etiquette has the effect of helping groups through potentially ticklish social situations, since it has the effect of both protecting participants from the unexpected indiscretion or blunder and providing them a way to regain balance once an untoward situation arises (Cahill 1990). Our ideas with regard to punctuality and tardiness fit into this discussion. While time has a functional dimension associated with the coordination of society, we also necessarily have buffers as regards understating how late we can be and what must be done to smooth over the situation when these boundaries have been transgressed. As we will see, it is into this context that mobile telephony has been placed.[15]

Mobile Communication and Microcoordination

Arguably, the greatest social consequence arising from the adoption of the mobile telephone is that it challenges mechanical timekeeping as a way of coordinating everyday activities. It is possible to say that the mobile telephone has completed the automobile revolution. Where the automobile allows flexible transportation, up until the rise of mobile telephony there has been no similar improvement in the real-time ability to coordinate movements. When you were enroute, you were incommunicado. The mobile telephone completes the circle.

Looking more carefully at this, we see that the use of time as a coordination medium relies on broad access to reliable devices to which people can refer when making agreements. That is to say, agreements are made with reference to a secondary system, namely, time and timekeeping devices. In addition, we have built up a superstructure of etiquette that surrounds the system of time-based coordination. This defines the tolerances allowed for tardiness and the contours of power associated with who waits for whom.

With increasing personal access to mobile telephony, we move away from the parallel interpretation of a common metering system and provide for direct contact between the partners who are coordinating their interactions. The mobile phone is competing with or perhaps supplementing the wristwatch as a way to coordinate social interaction. In a sense, mobile telephones allow us to cut out the "middleman." Rather than relying on a secondary system — which may not necessarily be synchronized — mobile telephony allows for direct interaction. This, in turn, provides flexibility and an efficiency that is unavailable with time-based coordination (Cooper *et al.* 2002; Townsend 2000).

The need for coordination has become more important as transport systems have grown and become pervasive. When looking at the role of telecommunication in contemporary society, it has been noted that transportation and communication are complementary systems. Beyond simply being a convenience, telephonic management has allowed for the coordination of activities that otherwise would demand extensive transportation (de Sola Pool 1982). The coordination of transportation and nomadic work has, with a few exceptions, such the police and taxis (Manning 1996), been based on interaction from geographically fixed nodes, such as telephone booths and the home phone (de Sola Pool 1977; Gillespie 1992). In this way the activity of transportation and its coordination have generally been separated. However, with the rise of truly mobile telephony, these two systems are being reintegrated and made more efficient (Katz 1999; Ling and Haddon 2001; Salomon 1985). Where we needed to coordinate movement from fixed nodes, now we can do it in real time (Townsend 2000; Cooper *et al.* 2002; Rheingold 2002).

It has already been noted that the niche of the mobile telephone is in the realm of small groups for microcoordination of activities. Microcoordination is the nuanced management of social interaction. Microcoordination can be seen in the redirection of trips that have already started, it can be seen in the iterative agreement as to when and where we can meet friends, and it can be seen, for example, in the ability to call ahead when we are late to an appointment.

Midcourse Adjustment

With the development of mobile telephony, we are open to far more finely graded coordination. Obviously, this opens up new possibilities (Katz 1999; Ling and Yttri 2002; Ling 2000c). At the most basic level this can be simple rebooking of agreements or the redirection of a trip that has already started.[16] It is here that the mobile telephone has the potential of increasing the efficiency of the transportation system (Lange 1993, p. 204; Cooper *et al.* 2002). Indeed, in work that I have done with Leslie Haddon, we have estimated that, on the whole, the mobile telephone saves more transportation than it generates (2001). This analysis showed that the opposite was true when considering the landline telephone. Much of the savings for mobile telephony comes as a result of the ability to redirect travel that has already begun. The device allows us to receive or request information along the way; thus, it short-circuits the need for meeting to exchange information. We can imagine that as programs change we must juggle our schedule and appointments. For example, when, due to the cancellation of a meeting, one parent becomes available to pick up the children at their after-school activities, the one can call or send a text message releasing the other partner — who may already be en route — from that assignment. The one might ask the other instead to go to the store for milk and bread. Further, if while in the store no whole milk is found, that person can contact the partner to find out if skim milk will do.[17]

We can see this in the comments of Ida, a parent, who noted that she uses the mobile telephone to send messages such as "Can you drive the youngest one to music lessons?" "Can you get him?" "Can you go to the store and buy milk?" Many of the same sentiments are seen in the comments of Anne:

> ***Anne:*** *That is pretty nice. If you are away from home and cannot reach somebody via the regular telephone, then you can call and leave a [SMS] message. So we use a lot of messages.*
>
> ***Interviewer:*** *What kind of messages do you send?*
>
> ***Anne:*** *It is if somebody is late, it can be if we need to buy something, if there is something important that he needs to bring home, if he needs to call somebody or if he has been home and has to give me a message. It is not like "Hi, I am doing fine, etc." It is something that we need.*

In many ways, Anne's final comment, "It is something that we need," is a type of credo we can associate with many aspects of microcoordination. That is, it is a type of instrumental function. The mobile telephone was seen as a way to assist in arranging the activities of one's children and other types of logistical issues, such as driving children to and from various free-time events.

> ***Ole Johan:*** *It is very nice to be available when you know that you have kids in the marching band and the soccer team, you know. And there are always messages regarding picking up and driving and not the least with the marching band. There can be driving for picking up things for the [annual benefit] flea market and things like that. And it is nice to be available, that you can get messages and such. It is things like that.*

Others described how the mobile telephone facilitates coordination their jobs and their private lives, blurring the boundary between these spheres (Mante-Meijer and van de Loo 1998).

> ***Nikolas:*** *Let's say that you are out shopping and suddenly remember that I have to buy something, let's say toilet paper or light bulbs. I park the car near the store. Then I can get a [job-related] telephone call there. It is like, I can go out of the store and talk with the person there and you can take a message and be available even though you are doing something in your private life, and that is very nice because working hours are quite long anyway, and you can sneak in things that you need privately during working hours. You are more free.*

Thus, we see here the use of the mobile telephone to adjust and rework a person's schedule and movements as various exigencies arise.

Midcourse adjustment is also seen in the adoption of SMS by persons with hearing difficulties (Bakken 2002). Before the development of SMS, these persons had to rely on extremely cumbersome systems to make appointments and organize social interaction. As with the wheelchair-bound person reported by Palen *et al.* in the previous chapter, the mobile telephone assists hearing-impaired persons when they have to readjust appointments (2001). If for some reason the hearing-impaired persons were unable to make an appointment and were marooned en route, they were unable to use the traditional telephone system to call ahead and to reschedule. This was particularly true if their counterpart was also hearing impaired. The use of SMS has revolutionized social interaction in this group because it allows the transmission and reception of written messages. In fundamental way, the mobile telephone has emancipated these persons and provided them greater freedom of movement.

Iterative Coordination

Another dimension of microcoordination that it is iterative. In this case, the mobile telephone is used to progressively refine an activity. Perhaps some friends are faced with a dynamic situation in which they need to progressively reach agreement as to when and where they are to meet. The friends can, for example, agree to "do something this weekend." That is, they agree to meet in general terms but need not specify the location or the exact time. Then, through progressively specific messages or calls, they can, in effect, zoom in on each other as the time approaches. Unlike the softening of schedules described in the next subsection, this type of coordination is an ongoing activity that takes place between when a general agreement to meet has been made and the details are progressively filled in. People call or send messages that give increasingly specific information as to where and when the meeting will take place. In addition, the place and the form of the meeting can be refined along the way. While it may have started as an open intention to meet, it becomes refined to meet on Friday after work. Next, a particular bar may be chosen, followed by a more precise time. The location can be shifted due to an unforeseen meeting late on Friday and finally there might be a series of messages updating the partners on their specific progress while en route.

A father, Ola, describes another variation of this type of coordination.

> ***Ola:*** *In the summer I work as a soccer coach for some 14-year-olds. And when you have 50 of them with all that that includes of this and that. Like yesterday, it started to rain a little and so you get 15–20 [calls] in the space of a very short time asking if there is going to be practice and a lot of things like that. So I could do without it, but I have to have it, you know.*

Rather than having to coordinate with a single other person, he reported the need to coordinate a situation that was in flux vis-à-vis a small platoon of teens.

A more dramatic version of the same phenomenon is the "swarming" described by Rheingold (2002; see also de Gournay and Smoreda 2003). Here, groups of persons follow the movements of public personalities, coordinate protest marches, or orchestrate the actions of others on a dynamic basis. This employs the idea of a telephone tree, with the dynamic nature of the mobile telephone, to update and coordinate the actions of many persons in near-real time.

The ability to contact others whenever and wherever adds a dynamic character to the interaction. We move away from the need to nail down a specific time or place. These elements can be determined in real time as the need arises.

Softening of Schedules[18]

The mobile telephone also has the ability to "soften" schedules, in that it adds slack to the more precise nature of time-based agreements (Ling and Yttri 2002). For example, if you are caught in a traffic jam, you can call ahead to your potentially frustrated meeting partners and advise them of the situation. In this way the agenda for the meeting can be juggled so as to allow the meeting to proceed while the person on the highway postpones their presentation a bit. Thus, mobile telephony can help to relax the scheduling of events, because calling ahead provides us with the opportunity to renegotiate arrangements. It helps our meeting partners to understand and navigate through a difficult situation, thus avoiding a breach of manners (Palen 1998; Blaise 2000).

The mobile telephone relaxes the implicit contracts around time. It softens the schedule. It is the tyranny of the schedule, and not necessarily the tyranny of time, that is at issue here. It is our sense of the need to respect our coordination with other people and institutions that is at play. In this way, mobile telephony plays into the courtesy of time and timekeeping discussed earlier.

As we have seen, the development of increasingly precise and accessible timekeeping has allowed for the rise of more precise, and perhaps more insistent, scheduling (Palen 1998). In one sense, the mobile telephone helps to push this even further, in the sense of the iterative coordination noted earlier. At another level, the mobile telephone helps to smooth over the more jagged effects of precise scheduling and coordinating.

This is particularly true with reference to the vagaries of automobile-based urban transport. When, for example, we have to meet expectations on several fronts (for example, working mothers), the mobile telephone can be used to assist in the transition from one role to another, i.e., from our working role to that of being a parent.

> ***Kari:*** *[I use the phone to] say that I will be late if something happens on the way home. Then I can call the kids and tell them that I am coming, that I am on the way. Before, I had to stop at a gas station to call if there was something.*

Kari's comments illustrate what Rakow and Navarro call "remote caregiving" (1993). In essence the device allows Kari a slight leniency because she can call ahead. We can also see this in the use of SMS messages. Analysis of more than 850 SMS messages[19] indicates that these type of coordinating messages are one of the major themes.

I will be there soon (Jeg kommer snart)

Come approx. five min late (*Kommer ca. fem min for sent*)

IM ON THE BUS NOW. (Original in English)

Am on the way home (*Er på vei hjem*)

In a traffic jam.lots of traffic.coming soon (*Står i kø.mye trafikk.kommer snart*)

Yeah I'm coming. (*Ja jeg kommer.*)

As with the situation reported by Kari, we see that the sender is en route but feels the need to inform his or her SMS partner of his or her status.

The use of the mobile telephone to advise others of one's status was seen as one of the most positive uses of the device. In the survey mentioned earlier, respondents were quite positive when responding to the statement "Using a mobile phone helps one notify others when you are late" (Figure 4.3). Analysis of the Europeanwide data[20] cited earlier shows that more than 92% of respondents agreed with this statement. Indeed almost three-fourths of those who thought that the mobile telephone was a disturbing influence in society agree that the device could be profitably used to call ahead. While there was a significant difference by age,[21] this was the only situation in which no gendered difference was found.[22]

These findings suggest that the mobile telephone provides a way to address a potentially difficult social situation. Indeed, the SMS messages shown earlier take on the form of being a type of courtesy afforded others.[23] Although the mobile telephone is often seen as a particularly barbaric device, the findings associated with this "softening" of schedules seem to indicate that, at least in some spheres of life, it is seen as a graceful way to resolve a practical problem.

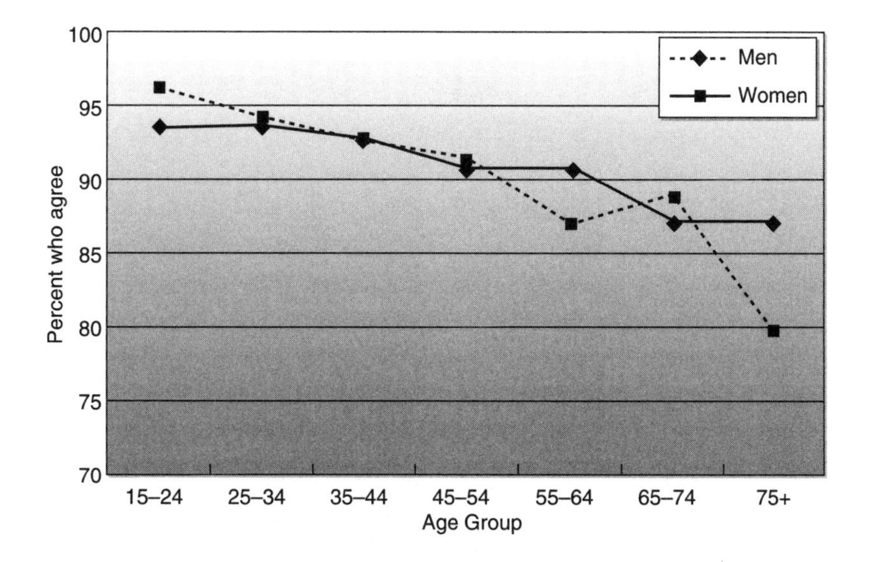

FIGURE 4.3 Percent of persons who agree with the statement "Using a mobile phone helps one notify others when you are late." From the Pan-European EURESCOM P903 study of mobile telephony and Internet use, 2000. (Note truncated scale.)

In this way, we come into the discussion of manners and etiquette noted earlier. Use of the mobile telephone to soften schedules can be seen as the type of self-reflexive situation. That is, the caller places him- or herself into the situation of the person called, and, based on this vicarious reversal of perspective, the caller can see that his or her own tardiness may occasion anxiety or frustration in the other. The motivation to call ahead is a way to defuse the potentially unpleasant feeling in the other. Thus, the behavior can be read as a self-reflective notion of how the caller would wish to be treated. This fits well into Duncan's notion that our use of courtesy and etiquette is an externalization of how we would like to be treated (Duncan 1970, p. 266).

Our common notions of what is courteous behavior are a social reaffirmation. It is through these rituals, in miniature, that we communicate our alignment with the broader social order and our expectation that the other is participating in the same order. In this way, the mobile telephone is a new tool we can use to deal more gracefully with complex scheduling situations. Interestingly, it can also be used to demonstrate power, in that our use of the mobile phone makes other, copresent people wait (B. Schwartz 1977).

Time-Based vs. Mobile-Based Coordination

Advantages of Mobile-Based Coordination

When considering the relationship between time-based and mobile-based coordination, we see several differences that make mobile-based coordination more flexible. First, one does not rely on an external metering system, but rather engages in an interactive process where the needs and situations of the partners can be progressively accommodated. Mobile coordination can increase the efficiency of planning meetings since the meetings can be renegotiated and redirected in real time (Townsend 2000; Ling and Haddon 2001; Cooper *et al.* 2002). The device obviates the need for contingency plans such as "If I don't show up by 6:00, go ahead without me." With access to mobile communication, we can quickly call to see if our meeting partner will make the date.

In addition, the partners to a meeting do not have to be geographically located in order to renegotiate their plans; they can do it literally "on the fly." Thus we move away from a type of linear conception of time in which meetings, social engagements, appointments, and assignments are fixed points at various time points to a situation in which these elements can, to some degree, be negotiated.

The mobile telephone can be used for the coordination of breaking events, such as the location of teenage parties on Friday night (Ling 2000c), gang fights (Lien and Haaland 1998), and the staging of social protests (such as those in Manila, Gothenburg, and Seattle) (Rheingold 2002), where slower centralized systems are at a disadvantage. Indeed, some claim that the success of the protesters in overthrowing the Estrata government in the Philippines was due partially to the ability to dynamically organize the actions of the protesters.[24]

Others have described what they call "swarming," which functions somewhat like a chain letter or a telephone tree, in that one person calls his or her closest friends to announce a breaking event. This small group in turn calls other friends, etc. In this way, the word about the event is quickly spread to a large number of individuals. This assumes, however, a generally one-way flow of information, i.e., a leader sending commands to his or her troop of protesters. It does not assume the type of interactive renegotiation of an agreement where all parties are on an equal footing.[25]

Limitations

Obviously, the mobile telephone has limitations as a coordination device. These include the number of people who can be coordinated, the need for large institutions to synchronize broad activities independent of personal need, and the relative fragility of mobile telephone systems.

If we consider only two persons, we can easily imagine the type of real-time interaction described earlier. The mobile telephone allows us to stop tracking time and to open ourselves for interpersonal communication in a more spontaneous way. However, when the number of persons to be coordinated increases, the complexity of the negations (and renegotiations) becomes far more profound. For example, if five friends agreed to meet on a Friday, the negotiation of the specific time and place would mean that one person has to call the other four and confirm the time. However, if one of the others cannot make it at the specific time or suggests another place to meet, then the whole round of calls has to be made again. When the number of persons goes over some threshold (perhaps 8–10 persons), it is much easier to use time as a common factor rather than an interactive process as with mobile telephony. This is particularly true if you are in the process of renegotiating a preexisting arrangement.

This point leads to the next problem with interactive coordination as practiced with the mobile telephone: that large institutions necessarily coordinate their activities via clock time. Meetings start, businesses open, trains, planes, and buses depart, and school lets out at particular times. If we consider an individual's interactions with various institutions, we see that the person is enmeshed in an array of organizations that schedule their activities by the clock, not via the dynamic agreements of the participants. Parents who have to pick up or deliver children to day care or who have to mail a critical package before the post office closes all know the terror of the clock. Within this system, the mobile telephone can, perhaps, smooth over some of the edges — particularly on the noninstitutional side of the equation. But it cannot stop the ongoing machinations of the various institutions. Here, time and timekeeping are in undisputed control. In this way the demands of institutions — or Mumford's "mechanical periodicity" — impinge on the organic forms of interaction seen in microsocial settings (1963, p. 198).

Another serious drawback with mobile communication is that the technology is neither as widespread nor as inexpensive nor as robust as the nearly ubiquitous time-based coordination. While some groups have nearly universal access to mobile telephony, others are less likely to own or use a mobile telephone. For example, about half of Bulgarians report owning a mobile telephone, but more than 80% of those in the United Kingdom, Italy, Norway, and Israel reported having one in December of 2001.[26] In Norway nearly 90% of young adult women aged 20–24 own and use a mobile phone on a daily basis, while only 18% of women aged 45–54 report the same (Figure 4.4).[27] Thus, there are dramatic differences in access to and use of the device. In addition, it costs money to send an SMS or to call. Finally, the technology itself is less robust than the alternative. The batteries can be dead, coverage may not always reach all corners of the caller's universe, and the devices do not tolerate the vagaries of the environment as well as modern clocks and watches.[28]

Indeed, the fallibility of the system has been institutionalized into a way of avoiding messages from unwanted quarters. Teens exploit this point when they wish to avoid being contacted by their parents. The strategic use of the on/off button and the heartfelt assertion that the battery was dead can serve as an excuse for being late.

Competition or Supplement?

It is perhaps too sweeping to say that mobile telephony is challenging the status of time as a coordination device. Indeed, analysis of SMS messages shows that time is often integrated into the coordination messages.

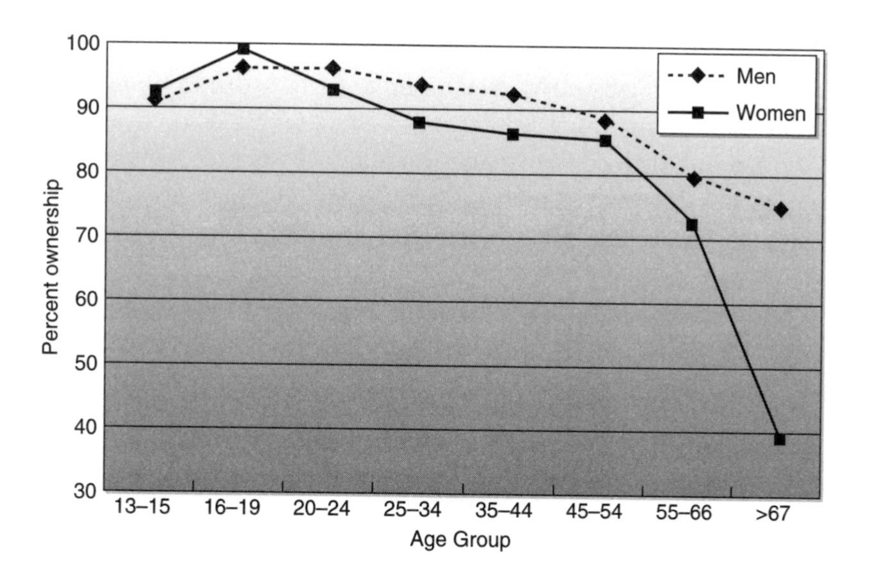

FIGURE 4.4 Percent of Norwegians reporting ownership of a mobile telephone. Telenor data from 2002. (Note truncated scale.)

> There is basketball practice outside19:30 (*Det skal være basketballtrening ute 19.30*)
>
> Is it ok. Then we have an agreement. 16:02 o'clock at my place. See ya, Me! (*De er greit. Da har vi en avtale. Klokka 16:02 hos meg. Hils. Meg!*)
>
> hi :) I can meet today and 17:00 o'clock see ya, Hug M___ (*hei :) ja jg kan møte dg kl 1700 snakkez! Koz M____*)
>
> Howdy⇒ Gallery tonight? 19 o'clock? I'll be there! Hug to you! I love you, Hugz⇒ (*Heisann⇒ Galleri i kveld? Kl. 19? I'll be there! Kos deg til da! Gla i du! Klemz⇒*)

In these messages, the sender uses both a mobile device and a reference to time in order to coordinate activities. In these four examples, the individuals seem to be involved in coordinating activities in the middle future. That is, they refer to activities that will happen some time soon but that have not yet begun. In general they are using rough time estimations, i.e., rounded off to the half hour.[29] In this way, they are different than the SMS messages associated with the discussion of "softening" time.

If we look at more close-grained coordination regarding activities that have already started, we can see time-based references of a different sort. Rather than referring to clock time, we see reference to elapsed time. In the following two examples, we see the SMS sender using the elapsed time to instruct or coordinate his or her correspondent as to when they will meet. The latter case here is indeed one of the stereotypical messages associated with SMS.

> I'm done w[ith] m[y] exam[ination] Meet me in 30 min[utes]. (*J er ferdig m xem Møt meg om 30 min*)

> Coming approx. five minutes late (*Kommer ca. fem min for sent*)

In these messages, particularly the latter, we see the tight interaction between time and mobile telephones in the coordination of social interaction. One system is supplemented with another. We see the underlying utility of time as a way to synchronize activities. However, this is a type of real-time coordination, in that the agreement can be updated as the terrain (or the traffic) becomes more obvious. Thus, coordination via the mobile telephone will not likely replace time-based agreements, but it will probably make them more organic in that they will better fit into the vagaries of everyday life. This means that in some situations we may be able to forget about time and indulge in interaction that is more spontaneous.

Conclusion

It seems that mobile telephony is potentially both a substitute for and a supplement to time as a basis for coordination. Mobile telephony allows for more interactive and nuanced coordination that does not necessarily rely on secondary systems. However, aside from the various types of swarming activities that arise, its application is in all likelihood limited to small-group situations.

While in some respects the use of the mobile telephone is jolting, that is generally because it is something new. However, it is in the process of pushing and nudging its way into a preexisting social context. It was not the case that we went from some sort of natural state of social interaction directly into real-time microcoordination via the mobile telephone. Rather, we have been so thoroughly enmeshed in an existing technical paradigm of time-based coordination that it has generally been taken for granted. In some respects, we have been in the position of being the fish that are vaguely aware of the water in which they are swimming. While we know of time and while we strive against it in planning our lives and coordinating with others, it is only when an alternative arises that time *sui generis* becomes obvious. In that situation, the assumptions regarding time and the morality we have developed become more focused.

This taken-for-granted situation becomes apparent with the adoption of the mobile telephone and provides us with an alternative way to coordinate our everyday lives. We see this when we can readjust plans with others to more effectively move through our daily chores. We see it when we can readjust the division of tasks on the fly, and we see it when we can use the time originally planned for one thing constructively doing another. We see the advantages when we are able to follow along with a rapidly changing situation in which we are involved, and we see this when we are tightening up a loosely structured agreement with a close friend. Finally, we see it when we are stuck in traffic and need to be someplace else.

It is in this context that a type of moral debate arises. How, when, and where do we use the mobile telephone to coordinate activity? As developed here, time and the respect for time has a well-entrenched position in society. We have relatively clear, though culturally varied, notions of punctuality and tardiness. The mobile telephone does not command the same position in society.

Use of the mobile telephone is often considered a frivolous break in the flow of a situation. As will be discussed in Chapter 6, it is seen as an abrupt break in the flux of a current situation. However, the mobile telephone is also a new tool that is playing with existing norms. I have noted that a large preponderance of people believe that the mobile telephone shines at its best when it allows us to tell others that we have been delayed and will be slightly late. In this respect, it is a new element in the arsenal of manners we have at our disposal. It is a way to smooth over the situation and avoid further harm.

Thus, at both a functional and a moral level, the ability to coordinate and microcoordinate via the mobile telephone relies on the same sense of social interaction as time and timekeeping, particularly when discussing small-group interactions. We are currently living through the phase in which the device is obvious to us. It is new and has not yet found its natural place. Currently we often see that while time keeps things from happening all at once, the mobile telephone seems to have the opposite effect. However, as with other technical developments, indeed as with time and timekeeping, it will fade into the haze of being taken for granted.

5

CHAPTER

The Mobile Telephone and Teens

Introduction

In the previous chapters, we have seen how the mobile telephone provides a sense of security and the microcoordination of everyday life. These themes often arise when discussing mobile telephony. Another theme is its adoption by adolescents and indeed the establishment of a culture of mobile telephony.

Many adults have the sense that the use of mobile telephony among teens is a thing apart. They experience teens as having a competence and a style of use that distinguishes them.

> *These teens see me as an amateur when I use my mobile telephone in relation to what they do because there is a lot that my daughter does that I don't understand at all, you know. And it is so fast. She says, "Look Dad." I say, "Ok," and by the time I have looked she is done. Then I ask her what she has done. I am completely stupid, you know, in relation to her. I don't understand any of this. Up to now you can say that I have called from A to B. (Thomas, father)*

While Thomas understands the basic communicative function of the mobile telephone, its advanced manipulation and perhaps, by extension, its symbolic position in his daughter's life are less understood.

We find similar discussions when considering teens and young adults in Japan (Hashimoto 2002) and Finland (Kasesniemi and Rautiainen 2002). The same is true in

Italy (Mante-Meijer *et al.* 2001), the United Kingdom (Harper 2003), the Philippines (Ellwood-Clayton 2003), and a host of other countries.

The widespread adoption of mobile telephony among teens is a source of comment, probably because it has happened so quickly. When considering the adoption among teens in Norway in 1997 (Figure 5.1), we find that almost no 13-year-olds reported owing a device. In addition, ownership was quite low among those under 18 years of age. In 1997 significantly more boys than girls had mobile telephones.

The situation had changed by 2001, when approximately 90% of the teens interviewed in a representative national sample owned a mobile telephone. The age differences had largely vanished, and, interestingly, the data shows that a significantly larger number of girls than boys had a mobile telephone. Earlier in the adoption process, the boys seemed to function as pioneer users. However, after the device was firmly placed into the everyday lives of the teens, that is, after the functionality was explored and its position was in place, it seems the girls made the

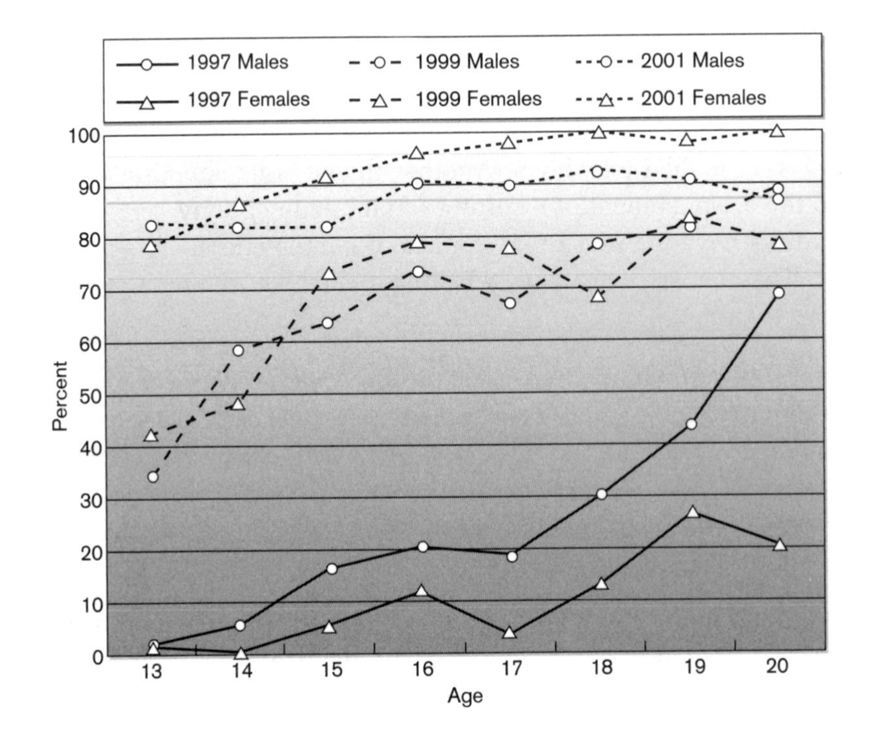

FIGURE 5.1 Trends in mobile phone ownership in Norway, by age and gender, 1997–2001.

mobile telephone into their own. Why has this taken place? What is it about mobile telephony that has appealed so strongly to such a wide spectrum of teens? In addition, what has this meant for interaction between parent and child and for interaction within the peer group?

Like all groups, teens use the mobile telephone to coordinate, and they use it as a way to provide a sense of security. Its use as a type of lifeline and its use in the coordination of everyday life are nothing if not functional and instrumental. Beyond the functional use of the mobile telephone, it has led to a different understanding of interaction and networking within the teen peer group. Teens in Scandinavia, Italy, Japan, and Korea have all adopted the mobile telephone, often to facilitate their social interaction. While many of the interests and activities are the same, the way in which they are organized is different. It allows for a type of anytime-anywhere-for-whatever-reason type of access to other members of the peer group (Ling and Yttri 2002). This means that the social network is more tightly bound together and that it is dynamic in its organization and location. Beyond this, the mobile telephone is also seen as a type of fashion accessory.

> **Interviewer:** *Why do you think that young people are so interested in mobile telephones?*
>
> **Kristian (23):** *Social status.*
>
> **Bjørn (22):** *Mobiles are a fashion thing.*
>
> **Harald (24):** *I agree with him. Mobiles have become trendy and hip. The more extreme your mobile, the cooler you are. It depends on which model you have.*
>
> **Kristian (25):** *Mobiles are status—the more expensive, the cooler you are.*
>
> **Anita (26):** *True.*
>
> **Sofie (24):** *Mobiles are like status.*

Adolescence is a period during which individuals develop their identity and sense of self-esteem. In this context it is possible to suggest that the adoption of the mobile telephone is not simply the action of an individual but, rather, of individuals aligning themselves with the peer culture in which they participate (Fine 1987, p. 133). This is perhaps more true of the mobile telephone than of other adolescent artefacts, such as clothing, since the mobile telephone is, in the first instance, an instrument for mediated communication. In addition, the cover, the type, and the functions are a symbolic form of communication. These dimensions indicate something about the owner. Finally, the very ownership of a mobile telephone indicates that the owner is socially connected.

From an interfamilial perspective, ownership of a mobile telephone means that teens are in control of their own channel of communication. It provides them with an independent link to others, and thus it represents a particularly central form of access. In addition, communication via the mobile telephone can be seen as being quasi-illicit. Calling during class period or late at night and the use of SMS for the exchange of sexual images are common examples. They point to the fact that teens are in the process of exploring sexuality and developing social interaction skills. In these ways, the mobile telephone plays into the peer group's role in the emancipation of the teen. It facilitates their learning how to manage quasi-illicit activities and in defining the boundary between what is proper and what is improper in various arenas.

Mobile telephony has become a part of adolescents' everyday life. The mobile telephone is used for the functional coordination of family and peer interaction; it is used by parents to allay their fear that their child is in trouble and out of reach; it is used to tie the peer group together; it has become a new point of orientation in the evaluation of status and fashion astuteness; it has allowed new forms of social intercourse and has become an element—and indeed a point of contention—in the interaction between child and parent; it scares the dickens out of parents who fear losing control of their children; and it is a midwife to the eventual emancipation of the children. In sum, it has become a part of teen life and the emancipation process.

Child/Adolescent Development and the Adoption of Telephony

A part of the issue surrounding the adoption and use of a mobile telephone is our understanding of general telephony. Telephone use is a background condition for teens. Indeed, it has been an integral part of preteen and teen life (Aronson 1971; Claisse and Rowe 1987; Kellner 1977; Lohan 1997; Mayer 1971; Veach 1981). However, learning about the telephone is a complex process. Young children have only marginal understanding of how to manipulate the device and how to manage talk over the telephone. How do we learn about the use of the telephone and the social interaction that is implied by its use? Insight into these issues will help to understand the adoption of the mobile telephone by teens.

Use of the telephone, not just the mobile telephone, implies physical access, the mastery of certain motor skills, psychological development, and linguistic competence with which to handle the particular situations presented by a telephone conversation. It is through understanding of the salience of these issues that we can understand the place of mobile telephony in teens' lives.

Norwegian data show that during adolescence there is a dramatic increase in the time spent on all types of telephony (Figure 5.2).[1] In the period between approximately age 9 and age 20, the mean reported number of minutes spent on the telephone almost triples, going from just over 10 to over 29 minutes per day. Young adults report relatively intense use of the telephone for personal interaction. This falls slowly to about 20 minutes a day for mature, preretirement adults. After retirement, there is a slight rise to about 22 minutes a day. The increase in use between ages 9 and 20, then, is the most dramatic transition in a person's telephonic life. This transition in telephone use takes place during the period in which we go from being generally within the sphere of our parents' home to being an emancipated individual.

How do we learn to use the telephone? The telephone is a fixture in children's lives from before they are social actors of any import. The telephone is held up to the baby's ear so that she or he can hear the voice of a distant relative. Small children are encouraged and cajoled into talking with their grandparents over the telephone

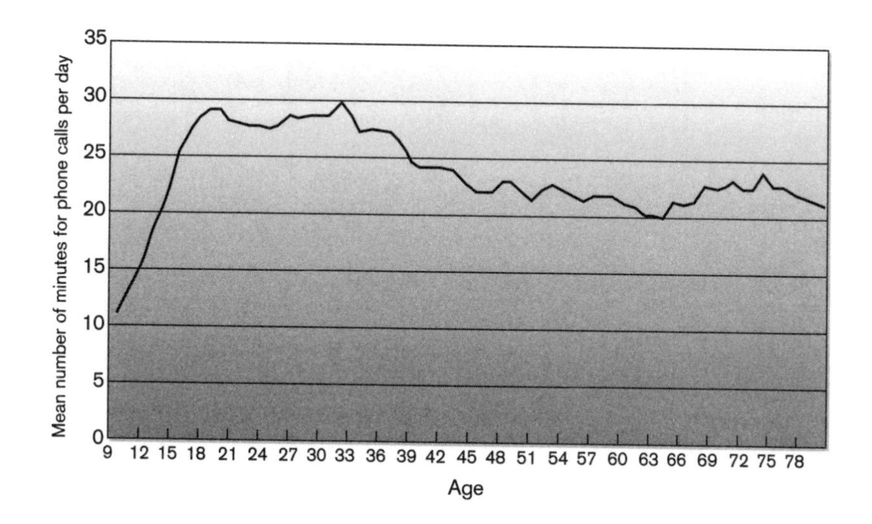

FIGURE 5.2 Mean number of minutes per day on the phone, by age, Norway, 2001.

before they really have a sense of how the telephone works, indeed before they really have a fixed sense of who these physically remote relatives are. Somewhat later in life, a child's use of the traditional telephone is prefigured by having access, a purpose for calling, familiarity with the telephone, and finally mastering certain aspects of voice modulation (Veach 1981).

Physical access is simply being able to reach the telephone, obtaining permission to use it, and being able to dial a number. Purpose vis-à-vis the telephone is, of course, having the need to talk to another. Regardless of how this is defined, the need must be formulated into the desire to use the telephone. After the child arrives at the telephone terminal, the child must understand the functioning of the device itself, i.e., entering the number and the meaning of the various tones and sounds that it produces (the dial tone, ringing, the busy signal, etc.).

Children do not necessarily have a comprehensive understanding of the technical system requirements of the system. Children must, for example, understand that they need to lift the handset before speaking. Holms describes an 8-year-old who said "hello" before lifting the handset of the telephone, which bespeaks a lack of technical understanding (Holmes 1981, p. 93). In her study of children's telephone use, Veach (1981, p. 99) reports another example. A 5-year-old child—referred to as "A"—tried to call a friend, Elijah. The transcription shows that "A" did not understand that Elijah was not on the phone while the phone was ringing. Thus, "A" attempted to greet Elijah before the telephonic circuit had been opened and before Elijah was available:[2]

A: *Elijah.*

(first ring)

A: *Elijah, Elijah, what . . .*

(ring)

A: *Elijah.*

(two rings)

AM: *Hello.*

A: *You know what I would like to say?*

While this example is charming in its naiveté, it also illustrates how much social and technical knowledge underlies our use of the telephone. It is clear that the caller was not completely aware of the way in which the system functions.

Children must also learn the special need to modulate their voice when speaking on the telephone. Given the background noise the telephone system and possible noise at the location of one's telephonic interlocutor, there is a need to speak slightly louder than in normal, face-to-face conversation and to articulate somewhat better.

All of these behaviours develop during childhood and, according to Veach, are largely in place for the traditional telephone by the time a child is 7 or 8 years of age. Children in her study who were younger than 7 had problems with various access issues. After that point, however, they were able to manage the motor skills and manipulation of the instrument. They understood how to dial the device and the connection between telephone numbers and persons.

Permission from parents was also commonly afforded after age 7, though Fine (1987) reports that this was not a carte blanche among the preadolescents in his analysis. In addition, they were able to manage irregular situations, i.e., unexpected noises produced by the telephone system, the various signals associated with dialing, etc.

From a psychological and psychosociological perspective there is the need to examine the role of cognitive development and socialization in our preparedness to use the telephone. There is the general discussion of developing a sense of the generalized other (Mead 1934, pp. 152–160). That is, the socialized individual can generally set him- or herself into and understand the situation of his or her interlocutor in order to manage a conversation. These skills are more important when considering telephonic interaction since we do not have visual access with our interlocutor.

Let's consider a young child speaking with a grandparent: The grandparent has a great deal of insight into the child's world, and thus some of the normal courtesies can be dispensed with. The themes of possible conversation are a type of common property that are already in existence beforehand and need not be built up from scratch, as with, for example, a fellow passenger on an airplane. The situation is more difficult for child–adult conversations when the two are strangers, since the adult has only broad understanding of the child's life and the child has no direct understanding of the adult's. Nor is there the social or linguistic competence on the part of the child to build up the necessary repertoire of small talk that allows the child to glide over difficult patches of conversation and to begin the process of making an acquaintance. The ability to deal with these issues in both face-to-face and telephonic conversation assumes that the individual can see the situation from the perspective of his or her interlocutor.

According to Piaget, as the child approaches and moves through adolescence, he or she moves from a more ego-centered to a decentered understanding of the world. In addition, the individual moves from a more concrete to a more abstract comprehension of relationships. Piaget suggests that by age 12–14 the individuals are able to deduce ideas about things that they have never seen. Piaget and those who follow his example suggest that this is essential, because it allows us to realize that others can have different perspectives. This, in turn, is important in the ability to take on various roles and to integrate various identity constructs (Piaget 1948; Brown 1990; Bush and Simmons 1981; Harter 1990; Savin-Williams and Berndt 1990).

Mead suggests a similar approach. He discusses the notion that as we mature we are increasingly able to take the role of the other, at least in some abstract way. This allows for insight into the perspective of the other and it allows individuals to gain a perspective on their own behavior. Mead goes further to suggest that the individual constructs a "generalized other" that is a type of mental summation of socialization (1925, pp. 268–275).

In addition to basic skills and psychological development, the child also needs to have a certain linguistic competence in order to use the telephone. To be sure, the telephone conversation presents the child with the need to comprehend several unique conversational and linguistic elements. These include the fact that the interlocutor is not physically present and thus cannot provide the normal visual cues and background information that makes the telephone conversation a special situation.

A normal conversation can be seen as an interaction between at least two individuals where, in its most basic form, there is an utterance on the part of one that is observed and understood by the other. The second person perceives, interprets, formulates a response, and then creates a "response" utterance. In the basic[3] formulation of the conversation, the first person then goes through a similar sequence.

The conversation just described is, however, only a reduced version of the interaction between two individuals. In addition to the central conversation, there is a secondary level of interaction that contains all the signaling needed for the management of the conversation. Other elements include groundings, i.e., the nods and small utterances that indicate that the listener is paying attention, the signals of our intention to speak, the clearance signals indicating that we are approaching the end of a conversation turn, the timing of speaker transition, and the management of topic transitions. There are also several levels of interproximate and interkinesic gestures[3] that help to facilitate communication.

In addition to the issues of the active management of a conversation, there are the issues of opening and closing a conversation, the timing between turns, changing topics, and repairing a conversation when unexpected exigencies arise. All of these issues have a particular salience when considering telephonic interaction (Ling 1998c).

Moving to the linguistic development of children, we can distinguish three areas of particular interest: (1) the various routines associated with a standard telephone conversation, i.e., the opening and the closing, (2) the speaker's timing, i.e., the pauses and overlaps between taking turns in a conversation, and (3) the management of topic transitions (Holmes 1981; Veach 1981).

Two particularly challenging points in the conversation are the opening and the closing. In these situations, especially the closing, there is a complex of signals passed between the persons in the conversation. With the telephonic opening there is a set of learned interactions during which the conversation partners establish contact, identify each other, and then elaborate the conversation (Saks *et al.* 1974).[4]

In a similar way, the closing is an achieved interaction between the conversation partners. The closing, however, is more complex since it comes when the conversation has run its course. The two partners may not, however, agree that the communication should be terminated. Thus, there is a set of signals that have to be exchanged indicating that the conversation is drawing to a close, e.g., increasingly long pauses between turns, various intonations, and the exchange of phrases such as "Okay" or even "Okay, bye." Nonetheless, the conversation can be reopened even after these preclosings have begun if one of the conversation partners comes upon another topic to be discussed, e.g., "Oh yeah, I forgot to ask you about..." When this topic is finished the partners can then also enter into a new closing negotiation (Holmes 1981; Schegloff and Saks 1973; Veach 1981, p. 113).

Veach found that children as young as 6 and 8 could routinely master the simpler opening routines on the telephone. She noted, however, that mastery of the closing sequence was common only among somewhat older children. The complete routinized version of the closing—i.e., the exhaustion of a topic followed by the preclosing exchange of, for example, "Okay," and then the formal exchange of good byes—was in place for 9- and 10-year-olds. However, it was only among the children 10–12 years old that Veach found the ability to vary the closing sequence, that is, to bring up new topics after the closing sequence had begun.[5] She also found that it was only when the children reached the age of 10 that they could adjust the opening and closing routines to different types of callers, i.e., adult vs. child, friend vs. stranger.

Turning now to the timing within a telephone conversation. Veach was interested in examining both the pauses between taking turns and the overlaps. She found that as the children got older, the pauses were both fewer and shorter in length. This speaks to the child's increasing mastery of telephonic conversation management. She also found that as the child grows older there were more cases of simultaneous speech. She notes that the number of speech overlaps actually increased among the oldest children in her sample. While we might expect overlaps in conversation to decrease as children gain competence, it is also easy to see that as they gain experience in conversation, they can begin to use the various grounding phrases discussed earlier to indicate their engagement in the discussion. Once children control the precision of timing, they can begin to use it to express meaning and emotion in their interactions. Thus, rather than indicating a lack of skills, it seems to underscore the growing mastery of telephonic interaction. Veach suggests that this occurs at about age 9–10 (1981, p. 276).

There is also the element of topic transition within a conversation. This complex issue requires a sense of the other's engagement, an overview of the flow within the conversation, and the mastery of the techniques for introducing new topics. Veach found that while the youngest children were not able to master topic transitions, children 10–12 years old were quite competent. The oldest children had the ability to see the conversation from the perspective of the other and also had assembled a repertoire of techniques with which to facilitate changing topics.

There are other aspects of sociolinguistic experience that children gain as they mature in their adolescent years. For example, there are quite strong gender-based linguistic characteristics. Women seem to be more accomplished in the introduction of topics (Fishman 1978), the use of rhetorical and factual questions to maintain a conversation and indicate interest, and forms of critique and interpretation when interacting (Treichler and Kramarae 1983). Adjusting the tempo of conversation, the transition between topics, the use of inexpressiveness, and perseverance in the maintenance of topics to control a conversation are also different for men and women (Sattle 1976). It has been found, for example, that mature women confirm their participation in a conversation more often than men and that they are also more likely to express interest through the manipulation of pauses and interjected linguistic grounding devices such as "mm" and "yeah" (Clark and Brennan 1991; Clark and Marshall 1981; Clark and Schaeffer, 1981; S. Duncan 1972; Johnstone *et al.* 1995; Kendon 1967; Saks *et al.* 1974). Many of these features are also seen in the use of the telephone (Rakow 1988, 1992).

Beyond the issues of gender there are status issues and the expectations based on dialect that are imparted linguistically, e.g., who can interrupt whom (Veach 1981, p. 192). There is also the management of intonation (Holmes 1981, p. 97).[6] These can be transition markers and signals of intimacy, emotion, or lack of understanding. Their management in a telephonic world is different from their management in a face-to-face world. Finally, there is the repair of miscues and problems in the conversation (Schegloff *et al.* 1977).

Thus, the linguistic dimensions of a telephone conversation develop throughout childhood and are mostly in place when children start into adolescence. In addition, however, they develop other aspects of their linguistic identity when they are well into adolescence.

These issues help us to understand the use of both traditional landline telephony and mobile telephony. The point here, of course, is that small children are not full-fledged telephone users.[7] Thus, while it is sometimes reported that young children have their own mobile telephones, they are not fully competent users until they have mastered a wide range of skills. These include mastery of the device as well as mastery of mediated social interaction. As children leave their preteen years, they are often quite competent telephone users. In addition, as they enter their teen years they have the motivation to use the telephone (and the mobile telephone) in order to organize their social interaction.

Adolescence and Emancipation in Contemporary Society

Beyond being the period when youngsters master telephonic interaction, adolescence is a period of social transition as well as physical maturation. One of the main tasks of adolescence is to progressively learn how to function outside the sphere of the family. Adolescents are asked to master a set of skills upon which they will rely in later life. These skills include, among other things, the mastery of their personal economy; interaction with various institutions and bureaucracies; dealings with friends, acquaintances, and even those with whom they are less disposed to be on friendly terms; the role of sex and sexuality; securing work and the expectations within the working world; and a sense of personal style and integrity. Beyond simply learning the formal expectations in these various realms, the adolescent must also learn the degree to which the formal rules need to be observed, the degree to which they can be flouted, and how to deal with the gray zones.

There are many transitions during this period. Children go from being a fixture in their parents' home to being emancipated (Schwartz and Merten 1967). The major elements of identity begin to fall into place. The individual's education, social and intimate relationships, career, and home are all in transition. In addition, there is the onset of puberty at or around this point, bringing with it a set of thoughts and impulses that push the attention of the child in new directions. Children have often begun to meet possible friends and partners, but the interactions are often inconclusive and perhaps often more confused than enlightened.

By the end of adolescence, many teens have moved out of their home of orientation into a transitory period before their establishment of a family. This young adult period is, in many ways, a type of extended adolescence (Frønes and Brusdal 2000). While not living at home or having independent income, the young adult often has not firmly established him- or herself in a family of procreation. This period may include the continuation of education, the indecisive pursuit of various jobs, and a type of sequential monogamy with a choice of partners.

In spite of, or perhaps because of, these stresses, adolescence is often a charmed phase of life. It is a well for contemporary culture. On the one hand it is a period of innocence, but at the same time teens often develop the friends and experiences that help carry them through life. Adolescence is intensely anticipated by preteens, and those who have passed into other, more staid periods of life often remember it fondly.

During adolescence, the child moves in an ever-expanding social circle. One of the critical social transitions for a child is the conclusion of primary school and the beginning of middle school (Brown 1990, p. 181). This transition, which comes at about age 13 in Norway, has the consequence of vastly expanding the child's social horizon. Before that point, the individual's social world is relatively small and local. One mother noted:

> *They don't have as big a range when they are 10 years old as when they are 17. They stay closer to home, and you have more of an overview of where they are. (Marta, mother)*

Along the same lines, a teen interviewee said:

> *How big a range does a 9-year-old have? It is, like, I was in the neighborhood, I was with a specific set of friends, you know. My parents could just call the parents of my best friend, and you were there and played or something. (Rita, 18)*

In most cases, when children go from primary to middle school they move into a world where their peers come from a much larger geographical area. The interaction with other students and teachers becomes more disjointed because teens usually move between various classes as opposed to remaining in one classroom (Harter 1990). The implication is that their social life becomes more complex (Robinson and Branchi 1997). The social network is much larger and the range of interaction grows. In a group interview, a father, Ola, spoke of the way in which the social horizon of the child expands.

> *These 12-year-olds, and 13- and 14-year-olds also, they are very active in that huge circle of acquaintances that they have with sports and visits. And when they begin with middle school, then they start different things and there are different confirmation things*[8] *and everything else. And they love to send messages to each other and to talk and to maintain contact, you know. They think that that is great. (Ola, father)*

In preindustrial society there was a stability from generation to generation. This meant that the knowledge and techniques of older generations were applicable to the situation of those who came after. It was the parents who could directly teach the child the ideas and techniques that were required.

By contrast, in industrialized society there is a flux. Children cannot expect to follow in the path of their parents, simply because the path has changed and is continuing to change. The types of jobs and the life experience of the adolescent are not reflected in the life situation of parents or grandparents. For example, the skills needed to use a PC or to send e-mail were not in the repertoire of the pre-WWII workforce. By the 1990s this was a common feature of many jobs. We also saw this gap in insight in the opening excerpt of this chapter. The father, Thomas, admitted that felt he was "completely stupid" vis-à-vis his daughter's competence with the mobile telephone. Thus, the knowledge and the perceptions of those who came before must necessarily be discounted to some degree. The degree to which they are discounted and the ideological baggage that must be dropped in this process is, of course, the stuff of many "coming-of-age" novels.

The fact that one generation does not have full insight into the skills needed by the subsequent generation means that institutions are necessary to fill in the areas outside the competence of the parents. Thus, there is formal education and other, quasi-formal institutions, such as summer camps, peer groups, sports teams, and irregular jobs, that provide the teen with necessary insights.

The age grading of the school system means that teens experience large portions of their time with a same-age peer group that takes a central role in their activities, their sense of identity, their consumption patterns, and their orientation (Hogan 1985). The influences of parents, schools, and other institutions are not simply focused directly on the individual; rather, these influences are moderated and understood through the lens of the peers. According to Sullivan (1953, p. 257), "The preadolescent begins to have useful experiences in social assessment and social organization. This begins with the relationship which the two-groups [dyads] come to have to larger social organization, the gang."

The experience with other peers is essential. On the one hand, children's relationships to adults provide a sense of an ordered social reality. At the same time the peer group provides them with the sense that they can modify social interactions, and thus these relationships provide mutual meaning (Youniss 1980; Harter 1990; Youniss and Smollar 1985; Savin-Williams and Berndt 1990; Giordano 1995). Adolescence and the peer group are social institutions that assist individuals in developing an identity. This, in turn, provides them with some of the ballast needed in later life. Peers provide self-esteem, reciprocal self-disclosure, emotional support, advice, and information. They allow the ability to be vulnerable among equals, sensitive to the needs of others, and generally, perhaps for the first time, to acquire insight into social interaction outside of the family. Peer groups are largely protective of their members; they can draw a symbolic boundary around themselves and resist the intrusion of others. Fine (1987, p. 126) calls this an *idioculture*. It can include a whole system of nicknames, jokes, styles of clothing, songs, artifacts, etc. For these reasons, adolescence is that period when the peer group and friends are most central (Rubin 1985). We see this in the unquenchable desire of teens to be together with and to communicate with their friends, and here we can see where mobile telephony fits into the picture.

At the same time that individuals find support in the peer group, they also experience teasing, gossip, and infighting. There can be chafing between the immediate peer group and the broader circle of friends (Giordano 1995). The peer culture's influence is also somewhat selective. While it has profound influence on the selection of cultural items such as slang and clothing, parents and the adult world are influential in areas such as career choices (Brittain 1963).

Thus, adolescence has developed into a type of social institution that helps to buffer the transition from the family of orientation to adulthood. It provides the individual with the space in which to take the elements of the old and the new and packs them into a comprehensive whole. In this way, teens are active in their own socialization. Glaser and Strauss have described this as the individual "shaping" the transition (1971, pp. 57–88). The notion is that the child, the parents, and institutions such as schools all have a say.

Elements in the Adoption of Mobile Telephony by Teens

The mobile telephone is a new element in the life of adolescents. It has sprung onto the scene with surprising speed. No generation of teens has had access to this type of technology, and thus access to mobile telephony means that teens—and their parents—are making up the rules of its use ad hoc. To put this into Silverstonian terms, while some teens are objectifying the device, incorporating it, and giving it a position in their everyday lives, their parents are still far back in the adoption process.

In mobile-intense societies such as those in Scandinavia, Italy, Japan, South Korea, and the Philippines, many teens have a mobile phone. How do we account for this popularity? What are the elements that led to this transition?

The development can be seen from several perspectives. Since it is a new twist on things and since it has implications for the way that teens exercise their growing freedom, the adoption of mobile telephony is the focus of comment. Indeed, we can say that there is a type of moral discussion regarding teens' adoption of the mobile telephone. This often focuses on use among preadolescents. Interviewees offer categorical and almost visceral responses. At this level of discussion, they did not need to justify or nuance their comments. Participants in Norwegian focus groups felt that it was simply wrong that younger teens and preadolescents owned a mobile telephone.

> *I have to admit that it is sick that a 10-year-old child has a mobile telephone. (Grethe, 19)*

> *That is completely wrong. [My son] is 13. It is unthinkable that he will have a mobile telephone before he starts in high school. (Kari, mother)*

> *I know of some small children with mobile telephones and I don't like that. (Karl, 19)*

It is interesting to note that both parents and older teens seem to share this attitude. This speaks to the speed with which adoption among teens has taken place.

Indeed, young adults, who grew up without mobile telephones, commented on the shift from the experience of their not-too-distant youth.

> *When I was 10 years old I didn't even know what a mobile telephone was. (Ivar, 19)*

> *The reason that we react is surely because we are not used to it, because it was not like that when I was young. (Dorthy, 23)*

> *My sister is 11 and she wants a mobile telephone, you know. And more of, at any rate, the boys in her class have a mobile telephone. I was there [at her school] to get a house key a while back during recess and ... suddenly five telephones rang at the same time. (Andre, 23)*

> *I think that people are too young when they are 10 years. I don't know, but maybe because I got my mobile telephone very late. (Grethe, 19)*

A slightly more grounded discussion questions what the teen will do with the mobile telephone. In this case, there is a type of "needs" test associated with use. Indeed, this is a type of mantra in the discussion of preadolescent and adolescent mobile telephone use. In the words of one informant, "It must be necessary; it has to cover a need." (Marta 17) Another, slightly older informant (Terese, young adult) said: "What are they going to do with it? I don't understand that." She went on to say:

> *I think, at any rate, that it is completely wrong that 12- and 13-year-olds will have a mobile telephone. I have a sister who will soon be 14, and the thing that she wants the most is a mobile telephone. And there are several of her friends that have one. I think it is completely wrong ... I understand that if she is going out for the evening and mom and dad want to get in touch with her and things like that, but then she can borrow their mobile telephone. And it is that she is going to have it with her at school and things like that. I simply don't see the point with that because I don't think that, in a way, they are able to control it. (Terese, 18)*

Other informants brought up the same theme.

> *There is no need for a 10-year-old, there should not be. I am completely convinced of that. (Kari, mother)*

> *I don't think that [my son] needs it ... Is it actually necessary that everybody needs to talk with each other all the time? (Joachim, father)*

Others gave—sometimes overblown—examples of younger children who had extravagantly expensive mobile telephones.

> *I want to say that if one buys a mobile telephone for a 13-year-old that costs 10,000 kr. (ca. $1200) that it is no longer for a need but it is more to mark status, to show off, and that is something else. It can be done in a lot of ways, and one way is just that, to buy an expensive mobile telephone, and it is obvious that it is something that is "in" and increases one's status among the gang. And so they cover a completely different need than calling when you see what the need is, and the way that it is a way to show off. (Arne, father)*

This discussion examines the "needs" argument by contrasting it with the use of the mobile telephone as a status object. There is also a redefinition of the needs issue, in that the "need" for safety and coordination is accepted, but not the use of the mobile telephone for status display. The need for calling is acceptable, but not the consumption of unnecessarily expensive items in order to do that. As we will see, it is indeed used for these practical issues. However, to limit the analysis of teen use to these two issues is to miss the point. The ownership, display, and symbolic use of the mobile telephone are—in many cultures—an essential part of being an adolescent. Just as the symbolism of the car goes far beyond simple functional transportation for many American teens, the mobile telephone is packed into a much broader symbolic universe.

Functional Uses of the Mobile Telephone Among Adolescents

Teens, like adults, note that the mobile telephone can provide security and coordination.[9] In terms similar to those used in Chapter 3, parents made this argument with reference to teens that had begun to have a broader range of movement.

> *I have a boy who is 17 and is in high school, and he has not gotten a mobile telephone yet but he can borrow one occasionally. But now I am thinking about a cheap one, one that is functional to have, because he is beginning to go out now and then and he goes out to the city. And it is not that he needs to call me or that I am going to call him, because I don't do that either. But there is ... if something happens or something. He was at something or other last summer, a conference at the university, and he missed the last subway from Sogn or something. And then he could have called and we could have come to get him then. We had planned that he would go together with*

someone on the subway, and they didn't make it. And they stood there. With things like that I think that it would be great to have a mobile telephone. (Arne, father) (emphasis added)

Another mother took up the same issue when she noted her concern for her daughter. This concern was, however, weighed against the desire for her daughter to be able to experience the world.

Marta (mother): *I have a 17-year-old, and the worst thing I know is when she goes downtown. I am so afraid, but I just have to accept this, you know. But it helps that she has a mobile telephone because she can call if something happens. It is not to control my daughter that she should take her mobile telephone when she goes out, but it is, ahh …*

Interviewer: *For her safety?*

Marta: *"If something happens, call home and we will come immediately!" you know. Because she needs to go out and experience Oslo. She has to learn about the world.*

Martha describes some of the basic ambivalence associated with the emancipation process. Allowing a child the freedom with which to explore the world involves the weighing of a difficult balance between the child's need for a safety net and the desire for the child to learn, perhaps haltingly, to operate as an independent individual. It is, of course, an unclear balance. The mother sees it proper that the daughter is moving out into the broader world. Nonetheless, she is still concerned for her child's safety and protection in certain situations (Savin-Williams and Berndt 1990). The mobile telephone provides a partial solution here, a type of umbilical cord between parent and child. It is a discrete link. It can be used if needed while still providing the child the freedom of movement that she or he desires.

In addition to the provision of safety, adolescents use the mobile telephone to coordinate everyday life (Ling and Yttri 2002). In the case of preadolescents, this is not a central need. As children become older and their schedules and interactions become more complex, it seems that the mobile telephone is more acceptable.[10] The mobile phone affords teens with access. In the words of one interviewee: "Without a mobile it is not as easy to keep in contact with what others are doing." Another said: "It is easier to be available with a mobile phone. It is also more personal than a house [landline] phone."

Coordination was seen as one of the major uses of the mobile telephone. It allows teens quick access to information on their peer group's whereabouts and thus allows them quick mobilization (Lien and Haaland 1998). This was also found in the group interviews.

> *I imagine that 75% [of my calls] are like that. You just wonder about where they are or if they are coming or what they are doing or things like that. They just call to hear what is happening. We call before school to find out if they have left home or after school to find out what they are doing after school. (Frank, 17)*

Coordination with friends intensifies during those periods set aside for social interaction. According to one informant (Erika, 17), "On a Friday there are a lot more text messages than on the Thursday because people are out and need to find out what is going on." Since the mobile telephone is the communication channel over which teens have best control, it is used for the normal microcoordination of their nightlife.

There is, however, a sense that the logistics of coordination have taken on new dimensions. As noted in the previous chapter, time-based coordination seems somewhat inflexible compared to the style of coordination used by teens. We see this in the following sequence.

> **Inger (17):** *If you have a mobile telephone, you can change plans along the way. You do not need to agree to meet either; you can just call whenever you want actually.*
>
> **Interviewer:** *But how do you make agreements?*
>
> **Inger:** *I don't know. You agree where and when you are going to meet, and if there is a change you say that you will meet another place, for example, if that is easier.*
>
> **Arne (17):** *I usually just make plans by calling [on the mobile telephone]. "What are you doing tonight?" "I do not know yet." "Ok, I will call you later."*
>
> **Interviewer:** *It is such that you call and ask if you can do something together?*
>
> **Arne: Yeah.** *For example, today, when I am here, I can just agree with my friends that I will call them when I am done. Then it is easier than planning what you are going to do [beforehand].*

These comments indicate that access to the mobile telephone has changed the way in which teens manage their appointments. Rather than making a fixed agreement that includes both the time and place where the appointment will be fulfilled, there is a basic indeterminacy in the agreements. Teens describe a system of outlining several alternative activities on, for example, a Friday evening. As the time approaches, these alternatives are progressively refined in content, and alternatives are played off against one another. Finally, the teens commit to one of them, but perhaps alter the time, place, or content of the activity in the period immediately before it begins. It could include elements of other alternatives. If that activity turns out to be a dud, then there are other alternatives.[11]

This form of organization also means that social interaction among teens can be privatized. They no longer need to meet up at the appropriate street corner or drop-in site to learn when and where things will take place. Rather, meetings can be arranged for wherever and whenever. According to Haddon (2000), social interaction is being pushed into the home, and in this way the home is becoming a more public place. Referring to work done by Wellman (1999), he notes: "Generally, social activities that in the past took place in public are increasingly taking place in the home, which is itself becoming more public, more open to outsiders." Indeed, working in Norway, Sletten finds that there has been a large increase in the number of teens who have visited another friend's home (Sletten 2000).

A related issue is the status of those who do not have mobile telephones. There is a concern that they fall outside the planning of the social group. The issue, however, seems to be more nuanced. For example, in Norway, some teens do not have a mobile telephone, either because their parents do not allow it or because they have a type of ideological opposition to its use. Further, given the limited financial wherewithal of teens and given the prepaid-subscription system (which will be further described later), some of those who have a mobile telephone do not have credit in their account. Lack of access can mean that they are excluded, since they do not receive the information on where and when the group is meeting. On the other hand, the ubiquity of the device means that they need only hang out with a friend who has a mobile telephone in order to gain access to peer group information secondhand.

At the familial level, coordination also arises when considering the children of divorced parents. It has been found in earlier work that communications between nonresident parents and their children can be meaningful for both the child and the parent (Castelain-Meunier 1997). In some instances, it has been reported that the nonresident parents have purchased a mobile telephone for their children. Sometimes the children can be quite young. In cases where the adults are not able to agree on the rules of contact, this creates a parallel communications channel that is, to some degree, outside the purview of the resident parent. At the same time, the mobile telephone allows parents to communicate and coordinate with their child without needing to go through the filtering of the expartner.

Interview data provide a glimpse into this issue, as seen in the comments of a divorced father talking of his 16- and 20-year-old sons.

> *My two oldest, they live with me every other week, and then it is good to get in touch with them because then I know where they are, if they are at [their mothers's] house or out. Then I know where they actually are if I want to get in touch with them. I can call home to her and say: "Let me know when they are coming." But if I am a little late, it is okay [with the mobile telephone]. Then I feel that I have a little better overview over where they are. At any rate I can get in touch with them. (Tom, father)*

Here we see the use of the mobile telephone in a family with perhaps special needs introduced by the divorce of the parents and the age of the children involved. There is a sense that it is acceptable for them to use the mobile telephone for this type of activity.

Symbolic Meaning of the Mobile Telephone

It is too simple to say, however, that the mobile telephone fulfills only functional needs for the adolescent. Beyond its functional capacities, and like other items of consumption, it is a symbol (Douglas and Isherwood 1979, pp. 122–131; Silverstone 1994). In many respects, the mobile telephone has become an icon for contemporary teens in many countries. The importance of icons and symbols is that their meaning is easily transferred and bequeaths status to the user (Duncan 1972).

The mobile telephone is a particularly powerful symbol for adolescents, with their emphasis on peer interaction. It shows that they are accessible and in demand. Within economic constraints, it allows expressive integration and enables them to participate in "gifting," as seen in the sending and receiving of telephone calls and text messages (Johnsen 2000; Taylor and Harper forthcoming). The mobile telephone indicates that the individual has reached a certain level of economic wherewithal and perhaps a level of technical competence (Ling 1998a). The mobile telephone has impact on the breadth and depth of a teen's social network, and it allows for the quantification of popularity as seen in the number of names recorded in the device and the number of messages and calls received. Finally, ownership of the "correct" type of mobile telephone shows that teens are aware of the current fashion and that they are active in the creation and maintenance of their own identity (Ling 2001b; Fortunati 2003b).

The object itself is invested with meaning, and thus it is seen as a way for preadolescents to obtain the signs and symbols of the adolescent world. It is also seen as a way for the adolescent to get a foot into the adult world. That is, the mobile phone allows a type of presocialization. It is the adoption of the outward form of the next stage in their lives.

The symbolic value of the mobile telephone often precedes its actual possession. While there is a growing "functional" need for the mobile telephone as adolescents move into the midteen years, it is also seen as an object of desire by those who are still preadolescents. In the words of one respondent:

> *When you are in primary school, it is probably extra cool [to have a mobile telephone] because then you are so big. (Emma, mother)*

This desire is a type of presocialization in which children are starting to prepare for their coming role as a quasi-emancipated adolescent.

During adolescence, peer group membership trades on quickly rising and equally quickly disappearing trends (Fine 1981; Lynne 2000). Today's fashion is out next year or next week. This year's argot is gone, seemingly before other generations realize that it has arrived on the scene. In this way, the mobile telephone is a type of fashion item. That is, it has become an artifact that is easily interpreted and that bequeaths status to the user (Duncan 1970, 1972). Among teens the mobile telephone helps to express a sense of form and style—not only in the color and model of the device, but also in the type of ringing sound it has and the logo on the screen, the language used to describe it, and the lingo used in SMS messages.

> ***Martin (23):*** *The mobile phone is like clothes ...*
>
> ***Carlos (25):*** *If you have a Nokia you are cool; if you have a Motorola or a Sony-Ericsson you're a business guy ...*
>
> ***Anders (22):*** *If you don't have a mobile, you are out of it! The model has a lot to say, you know. A Philips "Fizz" from 1995 is nothing that you show off.*[12]
>
> ***Harald (24):*** *I think that blocks of cement are cool.*
>
> ***Carlos (25):*** *I am proud of my Motorola Timeport.*
>
> ***Peter (24):*** *If you got a Nokia, you are one of the herd; if you have something else, you will soon buy a Nokia.*

Here we can see the teens describing the cachet of the various mobile phones. A Sony-Ericsson is for businesspeople; the ironic description of the older Philips as a "block of cement" shows that it is out of fashion. Carlos asserts the status of his Motorola, and Peter describes the widespread popularity of Nokia. There are various takes on the status afforded with the different types of phones. All of them can be used to call, and all—with the possible exception of the older Philips—can be used to send SMS messages. Nonetheless, the teens are experienced at placing the different brands into a context of style and fashion.

To study fashion and personal display, as seen in the use and ownership of a mobile telephone, is to study individual intention (Davis 1985; Dichter 1985; McCracken 1988). Cunningham and Lab (1991, p. 5) note:

> *Material objects, such as clothing, help to substantiate and give concrete cultural meaning to individuals. They are the media through which cultural ideas flow. That is, clothing helps to substantiate the manner in which we order our world of cultural categories, such as class, status, gender, and age, and express cultural principles, such as the values, beliefs, and ideas which we hold regarding our world.*

Fashion also reflects our personality and group identity, i.e., gender, role, occupation, economic status, and political beliefs. Clothing, fashion, and the display of various artifacts can also be central in various cultural rituals, such as marriage and graduation. Both in everyday life and in ritualized situations we can use props, costumes, and artifacts to express a sense of self and how we wish to be seen. This is the Goffmanian front stage, where we present a specific façade and where we try, as best we can, to suppress impressions that could contradict the intended effect (Goffman, 1959, pp. 111–112). We feel an integration between the subjectively imagined image and the actual façade. We engage in various strategies—"face-work" in Goffman's terminology—in order to beat off threats to our imagined façade. Poise and savoir-faire are drawn upon in case our façade slips (Goffman 1967, pp. 12–13). Cunningham and Lab implicitly include this when they note, "Clothing helps to define our identity by supplying cues and symbols that assist us in categorizing within the culture" (1991, p. 11). The mobile telephone can also be seen in these terms.

There is a shift in the issues to be considered when we move from the perspective of the person effecting a style to that of the observer. We fabricate a presentation of self and provide others with cues and symbols that help them place us into some context. The viewers[13] also have a role. Following Goffman, the viewers are a part of our ability to maintain a face (Goffman 1967, 31):

> *A person's performance of face-work, extended by his tacit agreement to help others perform theirs, represents his willingness to abide by the ground rules of social interaction. Here is the hallmark of socialization as an interactant. If he and others were not socialized in this way, interaction in most societies and most situations would be a much more hazardous thing for feelings and faces.*

Thus, beyond the intention of the individual, there is an interaction in the interpretation of fashion and style.[14] Sociologist Georg Simmel has provided a key analysis of the social dimensions of fashion (1971). He notes that fashion is a dynamic mixture of dimensions.[15] We find a blending of the longing for individual statement and the simultaneous and opposite desire for group identification. Simmel writes: "Two social tendencies are essential to the establishment of fashion, namely, the need of union on the one hand and the need of isolation on the other" (1971, p. 301). Further, he notes (p. 296):

> *Fashion is the imitation of a given example and satisfies the demand for social adaptation; it leads the individual upon the road which all travel, it furnishes a general condition, which resolves the conduct of every individual into mere example. At the same time it satisfies in no less degree the need for differentiation, the tendency towards dissimilarity, the desire for change and contrast on the one hand by a constant change of contents, which gives to the fashion of today an individual stamp as opposed to that of yesterday and of tomorrow, on the other hand because fashions differ for different classes—the fashions for the upper stratum are never identical with those of the lower; in fact, they are abandoned by the former as soon as the latter prepares to appropriate them. Thus, fashion represents nothing more than one of the many forms of life by aid of which we seek to combine in uniform spheres of activity the tendency towards social equalization with the desire for individual differentiation and change.*

Teens in general and their relationship to mobile telephones are indeed excellent examples of this. As we have seen, adolescence is a period in which the individual is intent on establishing a personal style. At the same time, adolescents use fashion and the display of various artifacts as a way to identify with peers (Lynne 2000).

Within adolescent culture, as in the broader culture, there is the social need to mark boundaries (Flugel 1950). Within contemporary adolescent culture, a small sampling of the possible groupings includes, for example, the socially conscious, the protesters, the debutants, the punks, and the athletes, in addition to ethnically and gender-based groups. In all of these cases, teens use clothing and other artifacts—including the mobile telephone—to mark group boundaries (Douglas and Isherwood 1979). It is in the adoption and use of these common symbols that they indicate allegiance to the ethic of the group, and in turn the items themselves become an embodiment of "a joint spirit" (Simmel 1971, pp. 304–305).

> *From the fact that fashion as such can never be generally in vogue, the individual derives the satisfaction of knowing that as adopted by him it still represents something special and striking, while at the same time he feels inwardly supported by a set of persons who are striving for the same thing, not as in the case of other social satisfactions, by a set actually doing the same thing.*

Fashion in the form of clothing, idiom, and the ownership and display of various artifacts can be the basis for both inclusion and exclusion.[16] It can be a way to graphically—and, for example, in the case of tattooing, permanently—display allegiance. At the same time, the sense of what is fashionable can be used for excluding someone. Indeed, those who strive to adopt fashionable items but classically arrive too late never actually fulfill the full potential of the display. Indeed there is a collection of usually negative terms to describe these persons, such as *wannabe, upstart, arriviste,* or *nouveau riche*. By the time these persons have arrived on the scene, the cognoscenti, who surf closer to the edge, have moved on to other forms of expression. This describes an important issue in the Simmelian description of fashion, namely, that fashion exists in the tension between the popularized and the avant-garde. In order to be fashionable, we cannot be too far ahead or behind. Thus, consumption has to be done correctly.

Simmel (1971, p. 306) notes that fashion divides the past and the future:

> *Life according to fashion consists of a balancing of destruction and building up; its content acquires characteristics by destruction of an earlier form; it possesses a particular uniformity, in which the satisfying of the love of destruction and of the demand for positive elements can no longer be separated from each other.*

A particular fashion is in vogue for a period before becoming popularized in other—more déclassé—groups. Thus, fashion has the dynamics of a wave as it moves though the population. Simmel suggests that fashion occurs not at the lowest ebb but only after it has started to rise (1971, p. 302). However, as the wave reaches the apex, we are speaking of popularization rather than fashion.

Among teens we often see an overdone preoccupation with the manipulation and arrangement of the various props and facades (Harter 1990). This work is not always done successfully. As with any symbol, the consumption and display of a mobile telephone can be overdone. In the words of 17-year-old Erika, the inappropriate display of a mobile telephone "is still showing off, you know, like with cars, it is showing off." Another teen, Nora, notes, "There are a lot of people that use it like a status symbol." We can see in these comments that getting it right is a slippery business. One person's notion of good taste is another's definition of gaudiness.

Thus, Simmel's analysis presents us with two dimensions in which we can place fashion and, in the context of this discussion, the mobile telephone. On the one hand, there is tension between individual identity and group membership. On the other hand, we can consider the tension between the avant-garde and popularization.

Mobile telephony fits into this discussion. We saw that teens view mobile telephones in these terms. Harold's comment that "mobiles are have become trendy and hip. The more extreme your mobile, the cooler you are," in addition to his ironic description of large, unpopular devices as "blocks of cement," depicts the territory. In effect, he is describing the tension outlined by Simmel. To have status, you must know what is what. If the individual wants to indicate allegiance to the group, or indeed allegiance to being young and "with it," these are at least some of the terms of the discussion. Owning the correct mobile telephone is a way of confirming the correct participation in youth culture.

The mobile telephone is one of the ways teens develop and enforce group boundaries. We could see this in the earlier discussion of the Philips Fizz. The particular model or type of mobile phone is used to define a sense of group membership. The symbols, artifacts, slang, and irony of mobile telephony used within a group represent a way in which the group separates itself from others and creates a type of in-group solidarity. Those who are not in the group may not understand the distinctions among artifacts or the drift of the irony. Outsiders will not be able to use the slang with the same precision or to the same effect. This nuanced ability to make almost imperceptible distinctions based on the display of costume and use of lingo is a characteristic of adolescence (Lynne 2000). Indeed adolescence, with its focus on distinguishing itself from all that represents earlier generations, is one of the great motors of symbolic fashion, be it in terms of artifacts or in terms of slang.

In addition to enabling us to display our competence vis-à-vis fashion, the mobile telephone also allows us to mark status in more quantifiable ways. Indeed, teens can advertise their popularity through bragging about how many messages they have received during a typical day.

Another quantifiable measurement of popularity is how many names they have registered in their telephone. Teens often note that the memory space on their devices is full and that when they meet new people they are put into the difficult situation of having to erase the name of another person in order to make room for their new acquaintance.

The material in Figure 5.3 shows that among a sample of the Norwegian population, teens and young adults reported the largest number of names recorded in the directories of their mobile phones. There were significant age-based differences in the number of names registered in the mobile telephone.[17] The teens here reported an average of 98 names recorded in their mobile telephone. Those in the 20–24 age group reported 106. Those in the 55–66 age group reported only about 20, and those over 67 reported only 9. In general, men had a significantly larger number of names than women.[18]

As we can see in the following citation, it is important to be able to report a large number of names. It is a sign that the teen is successful socially.

Moderator: *How many names do you have in your name register?*

Nina (18): *Full.*

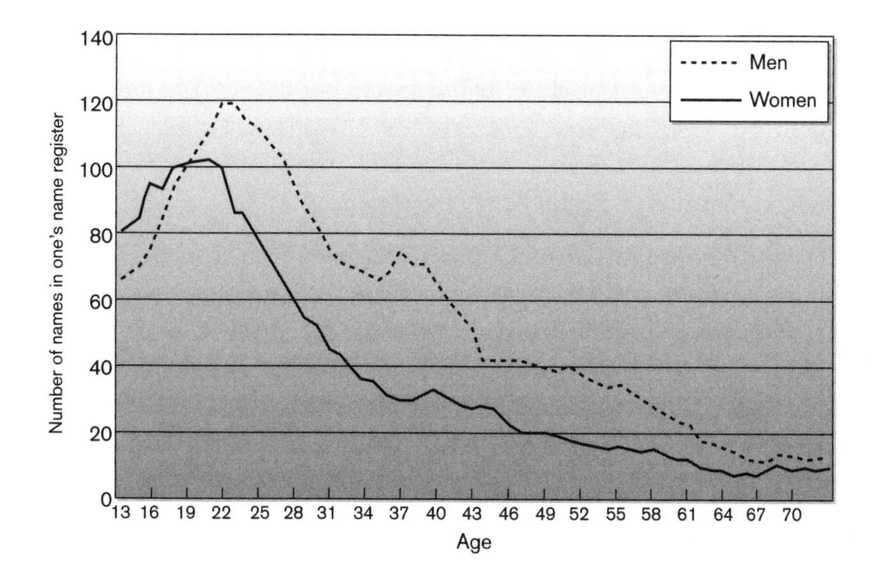

FIGURE 5.3 Number of names in an individual's mobile phone directory, by age and gender (7-year running average, Norway).

Moderator: *Full? In your telephone, how many are there?*

Nina: *100.*

Moderator: *100?*

Rune (15): *Mine isn't all the way full, I have about 80.*

Arne (17): *I have 100 or so, about 140.*

Moderator: *How many do you have?*

Oda (18): *I have 99 or so.*

Inger (17): *My telephone only has room for 50 or so.*

We get the sense that the number of names is one of the currencies associated with popularity. Arne, for example, goes from an estimate of 100 to 140 in a single bound, and Inger is left behind in the competition for names, since she can report only "50 or so."[19]

This leads to the notion that the mobile telephone is used by younger callers to enforce the idea that they are popular. Other dimensions can include a type of faux stress, in terms of having to answer many SMS messages or of problems with too many voice mail messages.

Erika (17): *I have received seven or eight messages from him today, and so I have answered seven or eight messages. But that is not the way it is every day, you know. When I come home, then I often have a pile of text messages from the day, but it varies in relation to who you are in contact with and what day it is.*

Another, more competitive version of this is seen in the comments of a man who coached a teenage soccer team: "I had a team in the Norway cup [a weeklong soccer tournament] and there were four boys and they had an internal competition. They erased everything [in their mobile telephone name registers], the one that got the most girls' numbers won." Again, we see the device being used as a way to quantify popularity.

Social Networking via the Mobile Telephone

Another way of developing a sense of solidarity is through the actual use of the mobile telephone to send and receive messages and calls—though it may be for seemingly meaningless reasons. Beyond the actual content of the calls and messages, this form of social interaction helps the sender and receiver to develop a common frame of reference and a shared experience. It is in this way that internal slang develops, news is communicated, caregiving and nurturing are performed, and courting takes place. With the mobile telephone, teens are given the opportunity to develop these relationships as the need or desire arises. A call—or more commonly a text message—serves to refresh the contact. The messages serve to tie the group together through the development of a common history or narrative.

The mobile telephone is a singularly strong symbol for teens, given their emphasis on peer group interaction. Its active use not only shows that are they accessible and "in demand," but also it allows them to participate in a type of "gifting" economy where they send and receive text messages (Bakken 2002; Harper 2003; Johnsen 2000). This is observed in the words of Nils, a teen who noted, "If you get a good message or one that is cool, you often send it on." We see here a kind of gifting that helps to objectify the relationship (Berger and Kellner 1964). Connecting with others is the essence. Where harried adults may try to reduce their interaction with the outer world, teens wish to increase the interaction. For example, we see this in the comments of Bente (18): "If I get a text message, I am curious. I want to be included, so, like, if I am in the shower and I get a message, I, you know, have to read it. If I write a message and don't get a response immediately then, it is like, you know, ehhh ...."[20]

Using Europeanwide data, Smoreda and Thomas examined the use of mobile telephony and SMS in the maintenance of social networks (2001a). Their results indicate that these two forms of communication are nearly a proxy for face-to-face interaction with a person's social network. Smoreda and Thomas found that face-to-face interaction is generally focused on friends living nearby. To almost the same degree, voice mobile telephony and SMS are focused on the same local friends. Thus, there is immediacy in mobile interactions. They are not used to maintain the more remote social relations.

There is also a gendered dimension to this issue. Teen girls seem to be more fluent in the use of multiple communication channels when it comes to maintaining contact with local friends. Using the same data as analyzed by Smoreda and Thomas,[21] I found that the more friends a boy had, the more he reported face-to-face meetings.[22] However, one finds only a modest correlation between the number of friends and landline[23] or mobile telephonic contact.[24]

The situation is different for girls. Here we find that, as with the boys, the more friends reported by teen girls, the more they report face-to-face meetings.[25]

However, there is a much stronger correlation between the number of friends and landline telephonic contact,[26] mobile telephonic contact,[27] and SMS contact.[28] Thus, the girls report stronger interaction across a spectrum of media when compared to same-aged boys.

Data from the Young in Norway study point in somewhat the same direction. The analysis of the material shows that teens who are moderately heavy users of SMS are more likely to spend time with friends than those who use SMS less often.[29] The material also shows that moderately heavy SMS users are less lonely,[30] use the PC,[31] and tend to be girls.[32] The data also show, somewhat dismayingly, that they had a tendency to play hooky more often than other, more moderate users.[33] In their favor, however, they were less likely to think that school was a waste of time or boring.[34] Thus, SMS users are relatively socially integrated, engaged individuals. While having a few quirks, the data seem to say that SMS users are well incorporated into their peer groups.

Monetary Dimensions to Teens' Adoption of Mobile Telephony

No discussion of mobile telephony among adolescents would be complete without an examination of its economic aspects. Indeed, it is among teens that the largest portion of disposable income goes to the use of mobile telephones.

Two marketing developments have been important here. The first is the prepaid, or "pay-as-you-go," subscriptions, and the second is the subsidizing of mobile telephone handsets by network operators. Prepaid subscriptions have been available for many years in the context of landline telephony. This form of subscription was generally used in the case of those persons not considered creditworthy. In the late 1990s this system was applied to the use of mobile telephones.[35] The second element that encouraged teens' use of mobile telephony was access to inexpensive, operator-subsidized terminals.[36]

Prepaid subscriptions and subsidized terminals eliminated two important barriers to adolescent use and released a pent-up demand for the services. In particular, the use of prepaid subscriptions eliminated the concern of parents that their teens would runup impossibly large bills on their mobile telephones.

These developments have also meant that the mobile telephone is often used as an object lesson by parents with regard to how to manage a personal economy. That is, teens are often given the responsibility of paying for their telephone use. This has meant that teens are forced to economize in their use of the mobile telephone, something that in itself is a motivation for the use of text messages, since they are generally cheaper than voice calls.

The analysis in Figure 5.4 shows that about 70% of the younger teens use prepaid subscriptions,[37] that this form of subscription is far less common among young adults, and that among those in their mid-20s only about 30% of mobile telephone users still use prepaid subscriptions.

The data also indicate that as we go from the youngest users to those in their late teens and early 20s there is an increase in the percentage of persons who pay all the costs associated with mobile telephony (Figure 5.5). Younger teens reported that others in the home often shoulder the costs associated with their use of the mobile telephone, presumably their long-suffering parents. Moving up the age scale, we see that as a person matures and becomes more established in a career, it is more common, in Norway, for his or her employer to subsidize the cost of mobile telephony.[38]

The shift from pre- to postpaid subscriptions and the fact that teens shoulder more of their own telephony costs reflect the case that they increasingly are working and that they have their own income. In addition, the fact that one relies on the mobile telephone extensively during this life phase and the fact that postpaid subscriptions are more economical also play into this issue. In addition, prepaid

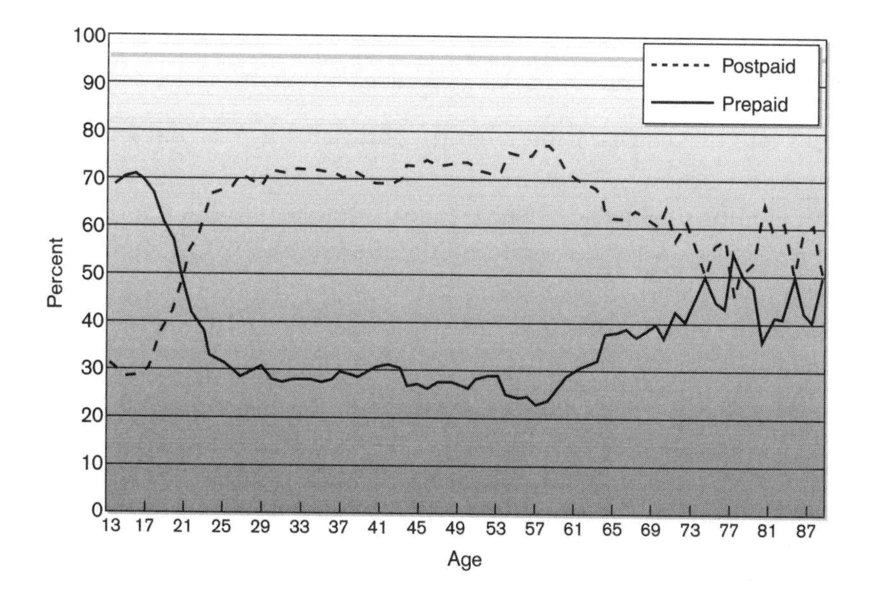

FIGURE 5.4 Percent of mobile phone users with prepaid vs. postpaid subscriptions, by age, Norway, 2002.

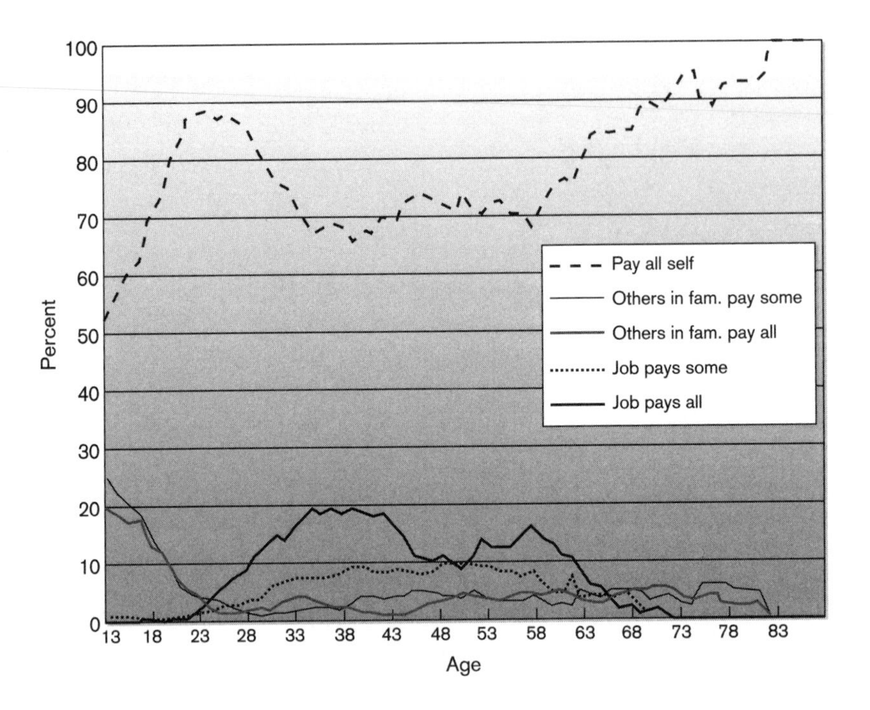

FIGURE 5.5 Source of payment for mobile phone services, by age, Norway, 2002. (1 = mean use for the sample)

subscriptions can be seen as being slightly childish, in that you are not responsible for your own economy. Thus, there are several factors that encourage the transition from pre- to postpaid subscriptions as teens move into the young adult phase.

When looking at the absolute amount mobile telephony is used (Figure 5.6), we see that young adults take the prize. When compared to the entire sample, teens report consuming about 25% less than the overall average mean. By contrast, young adults report using more than twice the mean for the overall sample. Young adults are often on their own and thus have only sporadic access to the subsidized telephone of their parents.[39] In addition, since they have often begun to have jobs but have not yet started on major investments, such as homes, expensive cars, and the like, they have access to more disposable income.

In addition, young adults are involved in a particularly nomadic period of their lives. They move frequently and have relatively irregular schedules, and so the establishment of a landline subscription makes less sense, both in terms of the effort to establish it and because it often does not fit their day-to-day lifestyle. Thus, young

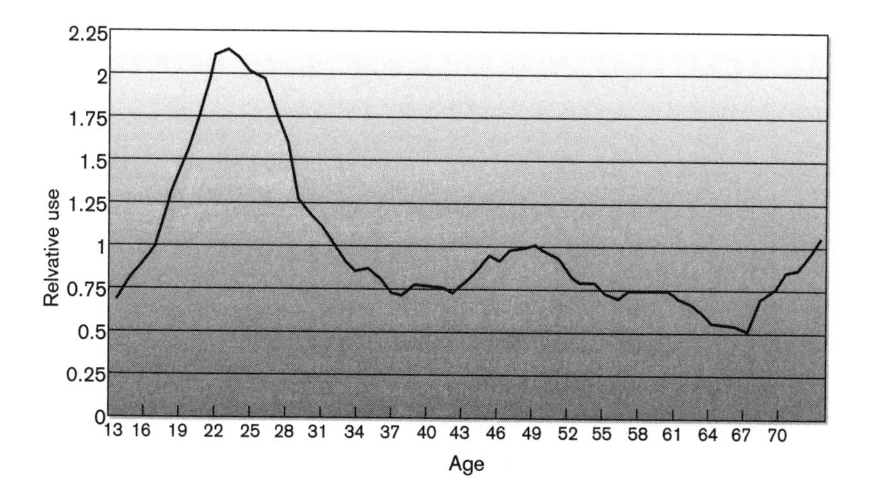

FIGURE 5.6 Relative use of mobile telephone services, by age, Norway, 2002.

emancipated adults have a very intense telephone use. Approximately one in four young adults has only a mobile telephone (Figure 5.7). No other group is as reliant on only mobile telephones as those in their early 20s.

Nonetheless, it is the teens who report spending the largest portion of their income for mobile telephone use. The data covering those between 13 and 24 (Figure 5.8) show that 13-year-olds report using about one-third of their disposable income for mobile telephony. Young teens' use is relatively intense, and at the same time they have the least access to income. Often they are dependent on weekly allowances from parents or other marginal forms of income. The percentage drops to under 10% for 18-year-olds and to about 6% for young adults. Thus, while young adults have the highest absolute consumption of mobile telephone services, in terms of money used their relatively robust economic situation means that the total impact on their personal economy is minimal.[40]

While the trends seem clear, there is a lot of discussion, uncertainty, and disagreement between teens and their parents concerning the payment for mobile telephony. While on the one hand teens can see payment as a badge of maturity, it can also be the nub of disagreement between parent and child (Ling 2000c; Ling 1998a). This comes through in the comments of 18-year-old Ida:

> **Interviewer:** *Has it been a topic of discussion with your parents that your telephone bill is too high?*

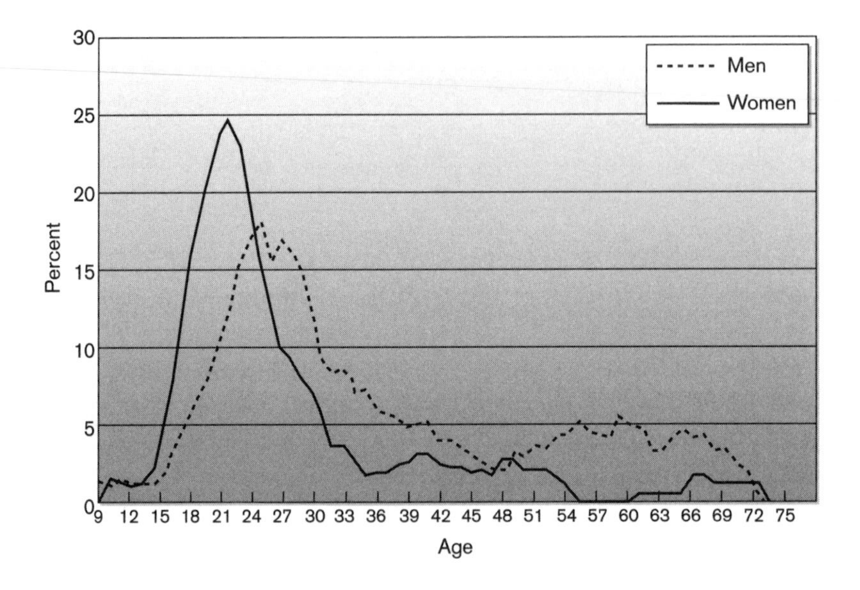

FIGURE 5.7 Percent of respondents who have a mobile phone only, by age and gender, Norway, 2002.

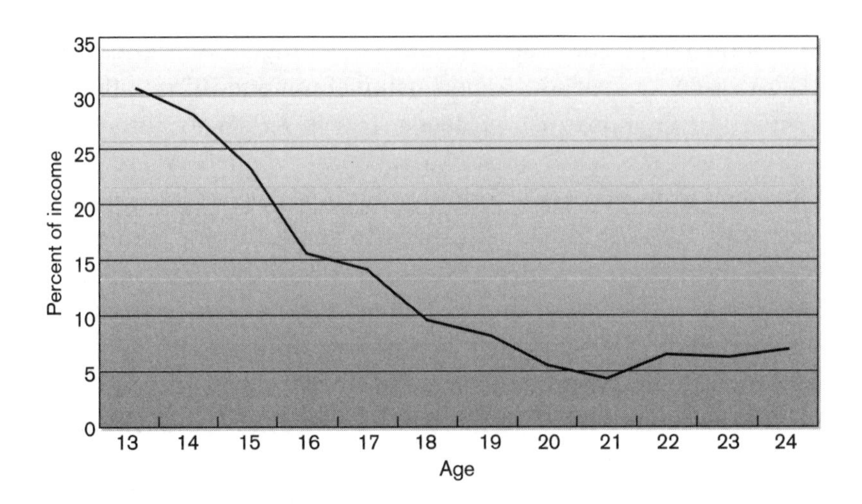

FIGURE 5.8 Percent of teen income spent on mobile telephony, by age, Norway, 2002.

Ida (18): *Yeah, more than once.*

Interviewer: *Have they tried to give you the responsibility for paying?*

Ida: *Yeah, they have tried, but they still pay.*

While Ida clearly asserts that she has managed to avoid paying for her use, her comments also betray the fact that it has been discussed, perhaps intensely, within the family. Insight into the negotiation process can be seen in the following dialogue between a mother and her 16-year-old son.

Grete (45, mother): *We have had some pretty big discussions about our last telephone bill. It grows and it grows and it grows. And we have tried to tell you. But, like it is, his friends, generally they have mobile telephones, each of his friends. When he calls a friend, it is a call to a mobile telephone. And if he doesn't call a mobile telephone, how is he going to get in touch? So we see our bill—it is incredible after a while.*[41]

Interviewer: *Have you made an agreement [on how to divide up the bill]?*

Grete: *Everything over 1500 Kr. ($200) he has to cover himself. He doesn't agree, but we have said that we will not cover anything over 1500. But it is specified, as I said. I see what it costs ... last time, and there were calls to mobile telephones for over 1000 Kr.*

Interviewer: *You think it is a little unfair?*

Fredrik (16, son): *No, not really.*

Interviewer: *You understand it a little?*

Fredrik: *Yeah, I understand it.*

Grete: *It is a bitter pill, isn't it?*

Fredrik: *Yeah.*

As opposed to Ida's ability to avoid payment for her telephone use, Fredrik has been forced to suffer the consequences of his telephone use. The sequence here indicates that he accepts this system, albeit begrudgingly. The interaction gives the sense that Fredrik is slowly moving toward the time when he will have to assume more personal and economic responsibility. At the same time, the empathy shown by his mother in the last exchange speaks to her understanding of the situation.

While on the one hand, teens sought to avoid the economic consequences of their telephone use, material also indicates that ownership of and payment for a mobile telephone are often seen as a way to gain "adult points." They are associated with the ability to earn money and control a budget. From the perspective of the teens, their ability to pay for the mobile telephone can be seen as a symbolic confirmation of their adulthood. We can observe the sensitivity of this issue in the following excerpt from a family interview in which a father suggests the need to help his daughter pay for her new mobile telephone subscription.

Walter (45, father): *But I have to say that if, like, with the new mobile telephone subscription, if it is so expensive, then I have to say that I could imagine that I will come with a contribution—that is obvious.*

Erin (15, daughter): *With the new mobile?*

Walter: *Yeah, if . . .*

Erin: *You are not going to pay for that regardless!*

Walter: *Yeah, I hear what you are saying.*

Erin: *Relax!*

Unlike the situation of Ida and Fredrik, Erin views her paying for her mobile telephony as a definite move away from the sphere of her parents. The sequence illustrates the unsettled and contrasting issues surrounding adolescence. Walter, her father, seems perhaps more willing to subsidize the telephone use of his daughter. For her part, Erin is assertive in her rejection of this seemingly unwanted interference in her personal affairs.

The mobile telephone has become a new locus among teens for discussing money and monetary issues. While children have limited economic wherewithal and often have an impulsive relationship to consumption, they are also often among the heaviest users of telephony (Ling 1998a; Mayer 1971). Thus, payment for mobile telephony is uncertain territory for both parents and children. Payment for telephony can be a point of tension, or it can be seen by teens as a declaration of freedom. On the one hand, there is pressure for the teens to pay for themselves; at the same time, it is an area in which parents see the need and perhaps the desire to contribute. It is an area where teens can assert their independence. At the same time parents use mobile telephone subscriptions as an object lesson in personal economy for their children.

Conclusion: Mobile Telephony and the Dance of Emancipation

We have seen that at several levels the mobile telephone opens up the discussion of the role of parents in the lives of their children. Even before the rise of mobile telephony, the issues of parental control vs. the freedom and emancipation of the children were complex. While the process generally flows in the direction of greater independence and responsibility on the part of the teens and less parental control, it is not a unidirectional movement. Rather, there is a series of episodes through which one or the other partner asserts his or her prerogative. A central part of the parents' role is to have an overview of their children's activities (Brown 1990, p. 179).

From the perspective of the parents, the goal is often seen as the need to control and channel their child. On the other hand, they express the desire that the child be a functional adult. The dilemma is not easy to resolve, and the mobile telephone underscores its arrival. Parents who were interviewed clearly thought that the youngest adolescents were still within their control. One mother, Marta, asked rhetorically:

> *Don't you decide for a child when they are 13 years old, what they should use their money for? Doesn't one do that?*

This issue is less clear later on. The midteen years seem to be the high-water mark of the transition. This is seen in the comments of informants in the "dual-career parent" group.

> **Marta (mother):** *But that day that you release them, that is a transition. There is a period when they are grown-ups and they are ... they can take care of themselves. But that middle period, that is dangerous.*
>
> **Moderator:** *When is the middle period?*
>
> **Kari (mother):** *When they begin high school (videregående skole), 16 years, from when they are 16 until they are 20.*

The mobile phone changes the power equation in this relationship (Manning 1996; Ling and Yttri 2003). Although the device allows for security and coordination, the child's ability to develop and maintain social contacts outside parents' purview is enhanced.

Teens in effect use the technology to manage interaction with their parents and with their peers. The caller ID function of a mobile telephone, for example, lets them avoid receiving calls from their parents when it is socially awkward. Since teens know who is calling, they need not answer. If confronted, teens have developed a repertoire of technically based excuses, such as the battery was dead or they had not heard it ringing.[42] Other, more advanced techniques are also available.

> ***Nina (18):*** *There are telephones where you can do it. Like, if a certain number calls, it goes right into the telephone voice mail. For example, if parents call, then it goes right into the voice mail.*
>
> ***Arne (17):*** *I do that.*
>
> ***Interviewer:*** *You do that?*
>
> ***Arne:*** *Yeah, when I am out on the weekend I do that.*
>
> ***Interviewer:*** *Who do you exclude?*
>
> ***Arne:*** *The family.*

So long as communication goes through the centralized family telephone, parents can maintain a certain, albeit oblique, overview of their children's social group and its interactions. The mobile telephone—and the mobile voice mail service that is usually combined with such subscriptions—is more private, and thus it allows adolescents to develop a parallel social world.

> *If I am not home and if I don't have a mobile telephone, then my parents would have been clear about all the people I hang out with. And if they [the friends] wanted to give me a message when I am not home but instead put it on the telephone voice mail, then they would have to be fast on their feet when thinking about what they want to say. When you have a mobile telephone, then you have private voice mail and a private telephone. (Erika 17)*

The way that parents and children play out their roles is the stuff of emancipation and also the stuff of adolescence. Diverse issues of control, coordination, security, peer group influence, and freedom are broached, and hopefully resolved, during these years. It is not a simple linear process; rather, it is episodic, and the adolescent's adoption of a mobile telephone is in many ways one of the defining episodes. It allows for safety, security, and coordination. At the same time, it can accelerate emancipation and loosen parental control. Obviously, this is seen as coming too early in the eyes of the parents and too late in the eyes of the teens, and therein lays the potential for disagreement, mistrust, and misunderstanding.

This disagreement seems to be at its height during the midteen years. Before that, the authority of the parents is most often intact; after that, the child has often established himself or herself as an independent individual.

The mobile telephone sets various dimensions of adolescence and youth into relief. Its use facilitates various aspects of social life, in that it provides for security and coordination. The important point here, however, is that in its symbolic mode, it also functions as a type of midwife for teens' transition into adulthood. It provides them with a form of symbolic capital, the opportunity to develop ideological positions, a form of group identity, and a demarcation between self and parents.

Indeed much of the reason that adolescents' adoption of the mobile telephone has caused such a stir is that it has butted into a set of preexisting and taken-for-granted notions of how things are arranged. It has raised questions of parental control, the way in which teens coordinate their activities, the ability of teens to perform as independent actors, teens' economic position in society, and a whole range of other issues.

To put this into the context of domestication, it seems that teens' use of mobile telephony is not a settled issue. Following Haddon's idea, the place of the mobile telephone is still being actively discussed and argued over. The moral and ideological furniture is not in place. The age at which children gain access, the conditions of access, the role of payment, and indeed the very need for a mobile telephone are still being agreed upon within the family. The different actors here all have different takes on the scene. The teens see it as the key to freedom (assuming mom and dad don't call at the wrong time). Simple access to the mobile phone is not enough, however. It must be the correct type, and it must be used and displayed in the correct way.

Parents are perhaps more wary. While there is often a sense that the mobile telephone has a place in the lives of their children, its exact place is difficult to pin down. If there is a functional need, then the case is easier to make. However, the more expressive uses of the mobile telephone are often more difficult to swallow. Of course their perspective on this changes as teens become more independent and assume responsibility for their own affairs, that is, as the emancipation of the children plays itself out.

6

CHAPTER

The Intrusive Nature of Mobile Telephony

Introduction

By now, many of us have been in a bus, elevator, restaurant, art gallery, or toilet and seen or heard, in Fortunati's words, a mobile telephone "erupting" onto the scene (Fortunati 2003). Indeed, the tale of a mobile telephone ringing at a funeral has become a modern urban legend. These sightings show just how widespread the technology has become, while the awkwardness of these situations points to the ways in which we are forced to reexamine, or perhaps reexert, our ideas of propriety.

Both qualitative and quantitative data suggest that the mobile telephone is seen as an invasive influence in public spaces. Qualitative data from across Europe (Klamer *et al.* 2000) as well as quantitative data (Mante-Meijer *et al.* 2001) support this point. Data show that when asked if they agree that "the mobile phone disturbs other people," almost two-thirds of the respondents in a European-wide survey either "tended to agree" or "agreed" with this statement. The data show, however, that there are differences between users and nonusers. As seen in Figure 6.1, about 60% of mobile telephone owners and more than 76% of the nonusers either tend to agree or agree with this statement.[1]

This, along with the use among teens, is one of the dimensions of mobile telephony that has perhaps been most thoroughly studied. The influence of mobile phones has been examined in restaurants (Ling 1997) and in terms of general attitudes

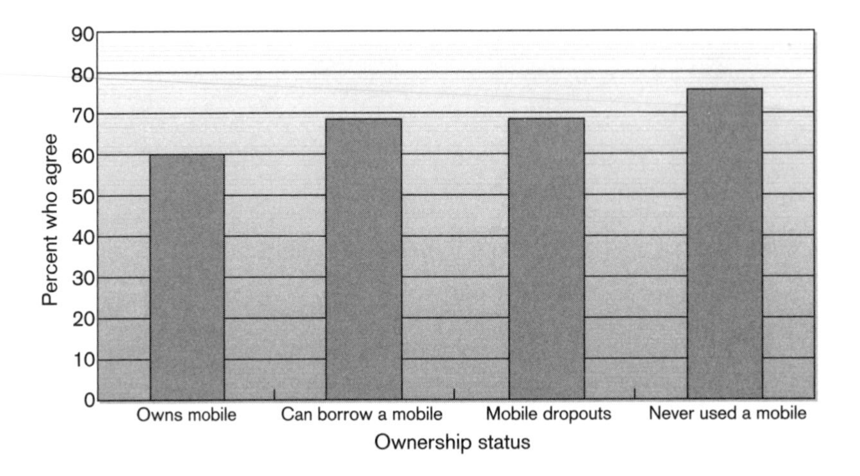

FIGURE 6.1 Percent of persons who agree with the statement "The mobile telephone disturbs other people."

toward the use of the device in public spaces (Ling *et al.* 2001; Mante-Meijer *et al.* 2001). Others have looked into the use of the mobile telephone in public spaces as a generator of urban myths (Håberg 1997) and the rise of mobile telephone etiquette (Nordal 2000), while Palen *et al.* (2001) have examined how new users perceived using mobile devices in public spaces. Finally, Love has studied the effects of personal space on the use of the mobile telephone (2001), and Murtagh has examined the use of mobile phones in trains (2002).

Data from interviews and focus groups indicate an almost visceral reaction to the inappropriate use of mobile telephones. The respondents were able to offer clearly formulated and well-rehearsed tirades against those who used mobile telephones in public. The fact that these formulations were readily at hand indicates that the provocation is not at the level of the individual; rather, it is at the social level. In addition, it is evidence of the debate at a somewhat broader social level over the norms of telephone use. It indicates that we are in the process of sorting out our collective sense of how to deal with this issue.

Respondents in focus groups were quick to point out various situations where they felt it was inappropriate to use mobile telephones. These included airports, stores, theatres, trains, buses, and various social functions and meetings.

> *Once [my husband] had [a mobile telephone] in town, and we were in town and were shopping, and right as we came out of a store the telephone rang. I thought it was very*

disgusting to stand there outside the store and talk, and he thought he was going to be so big and because I saw that. I think that was very disgusting.

Last week I was in a meeting and there were 10 or 12 at the meeting. Four of them had a mobile telephone. There was shuttle traffic out to the hallway. That is unacceptable.

I was at a Christmas party the other evening and there was a lady who flew in and out of the the circle of people around the Christmas tree [where children sing Christmas carols] three to four times. I cannot understand why you would have a mobile telephone at a Christmas party. And this with meetings: There are five men, each with his own mobile telephone. They ring just about continually; there should be a rule that you leave your mobile telephone and pager in the reception area.

There are three general areas where mobile telephony is problematic. Starting with the most broadly social, these include settings where there is an extensive set of norms governing behavior. Restaurants and theaters are examples. A second area is the complexity of using mobile telephones in the context of interpersonal interactions. The juggling between copresent and telephonic interaction is not easy. Finally, we have internal reactions to mobile telephone use, such as embarrassment.

Mobile Telephony in Settings with Heavy Normative Expectations

It is awful to see those who go around [using a mobile phone]. I could throw up! They use it even in restaurants. It looks so dumb! (Charles, 40)

The disruptive nature of mobile telephones is highly evident. Indeed some of the most visceral reactions to mobile telephony come in discussions of this topic. The device clashes with many social situations, particularly those governed by a heightened sense of normative expectations.

To understand this, it is important to see that buses, public spaces, and particularly restaurants are simultaneously public and private spaces (Lippman 1967; Mars and Nicod 1984, pp. 66–69; Silverstone and Haddon 1996). While a person is in the public domain, tables, booths, bus seats, and park benches become the "property" of the individual for the period of occupancy. To claim a territory, however, we must go through the rituals of establishing and agreeing upon illusory perimeters, or "symbolic fences" (Goffman 1971, p. 29; Gullestad 1984). We become quite accomplished at ignoring others who are in quite close proximity, through the use of fictive curtains.

There are often demarcations, based on architecture and the placement of furniture, that stabilize claims to space. These conventions allow for the erection of easily observable boundaries between individuals and parties on an ad hoc basis (Goffman 1971, pp. 33–34). In this way we temporarily identify where we belong and our expectations as to where other patrons have the right to intrude.

Within the stability there is also a flexible nature to these settings. In restaurants, where the use of manners and etiquette is more elaborate than in other public settings, tables can be divided or combined in order to accommodate varying numbers of participants. This notion of flexibility varies, however, with the class of the restaurant (Mars and Nicod 1984, p. 49). This minimizes the need for the patrons to readjust their "borders" with the entrance of other claimants in a type of "managed unreachability" (Gullestad 1984, p. 167; see also Haugen 1983). In other, more proletarian restaurants the customers often take the initiative to divide and redivide the space as the need arises. While this allows for a more concentrated complex of groups — something in which the proprietor is interested — it also means there is a need to adjust claims to territory, carry out cross-border raids when the need arises for extra chairs or condiments, and, in the worst case, possible wars of attrition when two parties lay claim to the same territory.[2]

In this context, a restaurant can be seen as a dynamic stage upon which our façade is displayed. It is a special situation where we are asked to combine etiquette and social finesse. It is where there is often a demand to celebrate social unity with our family, friends, or colleagues. At the same time there are many unexpected turns and twists that a restaurant visit can take. There is a well-prescribed set of rules, norms, and rituals that must be observed — the correct use of utensils, the way in which to eat, the topics available for conversation, etc. There are many possibilities for personal adventure — meeting that special someone in a singles bar, closing an important deal, savoring a specially cooked meal, celebrating a birthday or other significant transition, etc. Finally, there are many potential hazards — spilling the wine or the soup, making inappropriate remarks, the need to cover over the fact that one received a poorly cooked meal, having to deal with Uncle Fred after he has had too much to drink, meeting others who we had hoped to avoid in that situation, using the wrong knife and thus indicating unfamiliarity with "correct behavior," etc. (Chen 1990–91; Fine 1995; Giuffre and Williams 1994; Parker 1988).

The elaborate rules of etiquette indicate the importance of eating as a social performance (Jackson 1952, p. 325; Duncan 1970, pp. 266–269). In every culture, the rules of eating decorum are extensive and often include quite precise orderings of events and prescriptions of the exact placement of eating equipment, drink, and food (Mars and Nicod 1984, pp. 28–29; Rombauer and Becker 1964, pp. 9–25).

There is also room for interpretation, and unfortunately there is rarely perfect symmetry in the sense of decorum. The misuse of public space is in the eye of the beholder. If one party asserts his rights too boldly, the other side will feel that his or her status is being affronted. This rawness may indicate the lack of rituals with which to handle the situation. There are many difficult situations that are dispensed with through ceremonies that are almost taken for granted. The polite smile or laugh when we make a slight misstep shows our polish or poise. When we extend these rules to cover the use of mobile telephones also, then certain dimensions of their public use will recede into the background.

As discussed in Chapter 4, manners also indicate status and hierarchy. It is through our use of manners that we indicate the way in which we expect to be treated. Thus, holding the teacup with the little finger extended in the proper way is, in the words of Duncan, a "dramatization of the self" and a message to others that we expect a certain type of treatment (Duncan 1970, p. 266). Geertz also comments on the way in which manners indicate status. He notes that etiquette is a reciprocally built barrier that surrounds the individual. Etiquette protects the inner stability of the occupant, and as one goes up the social ladder, "the thicker [becomes] the wall of etiquette protecting the emotional life" (1972, p. 290; Gullestad 1992, p. 165). All of this attention to the process of eating, particularly when it is done in public, speaks to the importance of the event as well as to the potential for problems. It also speaks to the importance of reaffirming the social order (Cahill 1990, p. 392). All of these demands require that we become fluent in the maintenance of both our own façade and that of the situation itself.

A final aspect of a restaurant visit is an event with a clearly visible price tag. All participants know, with some clarity, the cost of eating out; it is listed for all to see on the menu. Implied in this is the assertion of status and the idea of gifting. Thus, to eat out is to give evidence of our position in society. To invite someone to a restaurant is in some respects to set a price on a social relationship. In a similar way, to accept such an invitation has the implication of becoming indebted to our host. Thus, untoward things that disturb the experience, such as the ringing of a mobile telephone, not only disrupts the work of maintaining façades, but also can depreciate the exchange between the dining partners. This sentiment was quite common among the informants, who noted, for example, "I have paid a lot of money for that meal" and "[Hearing a mobile phone in a restaurant] can make you feel like you've wasted money and made a bad choice about what you did that evening."

In sum, a restaurant meal is a context in which there is a large number of constraints and rules governing behavior. Other public settings, such as buses and park benches, have their share of rules governing interaction, but they are not as elaborate or as extensively specified as those governing dining. It is against this backdrop, not surprisingly that respondents in the focus groups cited the use of a mobile telephone in restaurants as a particularly galling social violation.

> *[Hearing a mobile telephone at a restaurant] breaks up the enclosed social circle that a restaurant offers. Seeing people communicate and be social at different tables is part of the restaurant experience.*

> *We can talk about common manners! When I was out at Mat & Vinhus [a local restaurant], it cost a lot of money to eat there, and the mobile telephone rang for somebody sitting behind us that had a lot of time to talk and talk. That is not good manners. I mean, he could have said, just one second, and gone out. (Sven, 45)*

> *I think it is repulsive to sit there and talk with people on a mobile telephone in a restaurant. (Bill, 39)*

Use of the mobile telephone can be seen as an affront to the decorum of a situation. As already discussed, courtesy is a way to buffer untoward behavior. Seeing the use of the mobile telephone as a threat to decorum means that it is among a category of behaviors that require special means to contain and defuse its potential for rudeness. In order not to be seen as being offensive, both the user and the others who are present need to develop and agree on correct behavior in a particular situation and on the ways to smooth over potential threats to this more or less common understanding.

The most basic objection to mobile telephones has to do with the sounds associated with their use.[3] When discussing personal space and the "modalities of violation," Goffman brings up "sound interference." In this case the violator fills up his or her accorded space, and then some. You can violate the territory of others by carrying out an encounter over a longer than proper distance and thus "obtruding" into the social space of others (Goffman 1971, pp. 33–34, 51).

In restaurants where it is the most difficult to make a long-term claim to space, e.g., cafeterias, it is easiest to use a mobile telephone. That is, a certain jockeying for personal space is taken for granted, and people are called upon to manage barriers most actively. In addition, the background noise may be of sufficient volume to cover the mobile telephone conversation. On the other hand, in those restaurants where the boundaries between parties are the easiest to define, e.g., more formal restaurants, where the individual is able to claim a table for a whole evening at the cost of a large bill, there is the assumption that the restaurant, and the other patrons, take responsibility for barrier maintenance.

Restaurant patrons have developed a set of responses with which they can cover unexpected sounds. The crashing of glasses or plates is parried with a smile; the interruption of a child is dealt with either through forbearance or a short scolding; a belch is either ignored or laughed at; the abrupt arrival of the waiter is tolerated; and the ringing of a traditional telephone is ignored, since it is not a signal of importance to the participants. The adroitness with which we can smooth over such social rough spots is important. Through the use of poise and savoir faire we minimize the damage of embarrassing incidents to the social situation. It seems, however, that the ringing of mobile telephones is not adequately routinized. Their ringing and the telephone conversations that follow cause difficulties in maintaining the ambience of the situation.

Beyond the ringing of the mobile telephone, respondents have noted that in the subsequent conversation those who used a mobile phone talk louder than in normal conversation. One noted, for example, that "people talk loud on telephones. Louder than usual, at least. That is as annoying as having a loud party next to your table." This brashness violates territories and makes it difficult to maintain the façade:

> *When you have a loud person talking into a [mobile] phone and you hear only half of the conversation, that disrupts the coziness of the restaurant feeling. (Hilde, 53)*

While many attribute the volume of talk to a desire for display or a sign of vulgarity, we can also examine this issue from the perspective how the technology plays into our form and style of interaction.

The telephone conversation is a speech event with distinguishing characteristics. Several of these break with the style and tone of the speech events that are common in face-to-face communication, where quite nuanced body language has several functions. Through our use of nods, glances, small sounds, and other gestures we indicate attention, the desire to speak, the desire to retain the floor, and pauses. We also use these devices to impart meaning and emphasis. All of these gestures are changed in a telephone conversation. Visual gestures are replaced by intonation and linguistic structure in "grounding" the conversation (Rutter 1987, pp. 105–126; Martin 1991, pp. 95–97; Duncan 1972; Saks *et al.* 1974).

Instead of relying on body language to control turn taking, pauses, emphasis, etc., we use what might be called verbal gestures. Tones such as "uh" replace the lack of eye contact that controls turn taking; phrases such as "ah ha" replace nodding and other signals of continued attention on the part of the listener; etc.

From the perspective of the mobile telephone user, it is perhaps not all that remarkable that we feel the need to speak loudly into the microphone. First, we are taught to speak loudly and clearly on the telephone, often long before we have a clear sense of how the device functions or even to whom it is that we are speaking for that matter. Young children often find themselves in the position of having an unfamiliar device pushed against their ear and having to suffer the request to "say hello to grammy." If they are capable of speech, they are further instructed to "speak up" so that the other person can hear (Veach 1981). In addition, telephone users adjust their speech to the audio situation. Thus, if the reception of our own mobile phone is spotty or if we have not managed to center the speaker of the phone over our ear, then given the low volume of the incoming sound, we will feel the need to ensure that our outgoing speech has the inertia to carry it through. Thus, if there is no background noise to cover the sound of our calls, we may very well be perceived as being inappropriately loud. Finally, we can become so engaged in the telephonic interaction that we forget our physical setting. Thus we can at least begin to understand some of the elements that lead to easily overheard conversations.

Mobile Telephony in Interpersonal Situations

Examining the use of mobile telephones in restaurants gives us insight into a situation in which the social context is guided by a particularly strong set of norms. These are in the form of the elaborate manners and etiquette that are associated with eating and formal dining. More generally we find that use of the mobile telephone in interpersonal situations introduces a new element into our interaction with others. The mobile telephone intrudes into the complex web of interactions, and it demands that they be rearranged. Understanding the complexity of these provides insight into why the mobile telephone is seen as intrusive.

The mundane nature of maintaining ties does not require much thought or planning. It is done through reading the interproximate and interkinesic signs given off by individuals.[4] Groups of individuals standing at a bus stop often arrange themselves into a small circle and exchange small talk, smiles, and glances with those in their circle. Accredited groups of individuals — particularly pairs — who are walking do this by arraying themselves shoulder to shoulder, exchanging glances, talking, or exchanging smiles or looks of concern, as the case may be. While the rules of etiquette are not nearly as prescribed as are, for example, those of eating in a good restaurant, there are nonetheless conventions governing our behavior that are taken for granted.

To the degree that it is made obvious, we see the management of these nuanced interpersonal situations where the status of the group is, in some minor way, threatened. We can see it, for example, when groups must negotiate doors, walk past oncoming persons, navigate down a crowded sidewalk, or decide to stop and look at something along the way. These circumstances require slightly more advanced interaction. Given these threats to the maintenance of colocated interaction, the partners must have at their disposal a set of devices that help them to maintain the status vis-à-vis each other (or within a larger group). There must be an openness to guide and be guided through situations that threaten the immediate status of the group. These can include taking the partner by the arm, a gentle bump, telling the partner to turn in one or another direction, pointing out a direction, etc.[5]

People walking together are able to maintain a running conversation, backchanneling verbal communication as well as the various forms of visual communication while threading through crowded sidewalks and narrow passageways. In this process they can play out the various presocialized ordering that underscore gender, age, and authority within the group.

Just as in our interaction with partners and groups of friends, we need to manage our interaction with persons who are outside the social group, in which case the purpose is in many respects the opposite. That is, instead of a repertoire of devices for maintaining the interaction, we need a repertoire of devices for marking boundaries and avoiding unnecessary interaction (Burgoon 1985, p. 361; Greenbaum and Rosenfeld 1978). Those who are not a part of the social group but who witness a face-to-face conversation between two partners have access to a slightly more abstract and limited version of both the verbal and nonverbal cues available to direct conversation partners. They may be able to see the way those who are more deeply involved in the interaction use body language and manipulate their voices either to enhance or to reduce the exclusiveness of the interaction. But they generally do not have access to the themes being discussed or the flow of the interaction.

The mobile telephone adds a new dimension to the management of this complex interpersonal interaction. In interpersonal situations, the use of the mobile telephone brings up the issue of how to include or exclude physically present persons in a conversation with a remote partner. This issue is made even more complex when secondary persons are present at both ends of the mobile phone conversation. These secondary people may be — to use Goffman's terms— *accredited* or *credentialed* members of the telephoner's circle or may simply be others who happen to be present at one or the other end of the mobile telephone conversation. The actual staging of the telephone call may demand that this host of secondary actors be either shielded from or perhaps included in various portions of the telephonic interaction.

Initiation of the Call and the Production of Social Partitions

It is at the initiation of the mobile telephone call that its invasiveness is most apparent. The ringing of the telephone is a quasi-dramatic shift in the flux of the situation. It introduces a new element into the action that we feel needs to be dealt with quickly and that changes the direction of the action, either literally in terms of the planning of an excursion (Ling and Haddon 2001) or in terms of the themes and staging.

The initiation of a telephone call can of course take two forms, we are called or we call. In the latter case, we have the chance to interact with those around us to agree on the need to place the call, on the contents of the call, and perhaps on the strategy or line that will be taken during the call. Thus, in some ways, the call is a type of collective interaction of the co-present accredited circle that culminates in the actual call.

Assuming, however, that we are receiving a mobile telephone call and not calling out, a different, more socially disruptive process is initiated. Our disengagement from a colocated interaction in order to answer a telephone is perhaps the trickiest point in a telephonic episode. We must somewhat quickly extract ourselves from the preexisting social situation and all the attendant front- and back-channel interactions. This is done through the use of various disengagement rituals.

The person receiving the call needs to draw on various gestures that exclude those in the immediate physical context. Simultaneously that person must place him or herself in a type of virtual context and initiate the greeting sequence for the telephone call (Saks *et al.* 1974; Schegloff and Saks 1973; Veach 1981). The copresent individuals who have not received the call must go through a parallel withdrawal procedure and thus partition off the telephonist from the local situation.

In the same way that the person taking the call goes through a process of local disengagement, other persons who are nearby must also take the change into account. Goffman's studied nonobservance is often a strategy used here (Goffman 1963, pp. 85–86). Goffman notes that these strategies allow the participants in a social setting to gloss over the situation and ignore these minor infractions to the order. In Goffman's terms, the various actors have a defensive stance, in that they wish to retain their own façade and to protect the façade of the others present (Goffman 1967, p. 14). All of this rearranging of the social furniture has to take place within a few seconds. Thus, there are many opportunities to offend sensibilities in this short time period.

In spite of the relative youth of mobile telephony, there are emerging norms regarding how to excuse yourself from preexisting social situations in order to take a mobile telephone call. These include various types of verbal cues, perhaps the mentioning of the person's calling as seen in a caller ID function, closing of the body language, and, if possible, complete retreat from the immediate area.

> **Observation:** A couple in their 30s entered a bookstore and approached a shelf with travel books, with the woman on the left and the man, shoulder to shoulder, on the right. They started to focus on a book that was at about eye level when the woman reached to pull a book off the shelf. At this point the mobile phone of the man emitted a single "peep." He dropped his glance from the shelf, and as he drew his phone out of his pocket he muttered the half sentence "It must be ..." By this time he was able to get the phone out, activate the call, and begin the greeting sequence. The woman accepted both the ringing and the half sentence without removing her glance from the shelf of books. The man, however, backed away from the shelf, assumed the closed posture of a mobile telephone conversation, and sought out space in some of the less trafficked portions of the store.

We see how the man and the woman encapsulate the episode through a dialogue of gestures. The man closed his body language and gaze. In a similar way, the woman dropped her interaction with the man and focused on the selection of a book. The use of gaze and posture and the gesture of the hand to their ear[6] underscore that we are engaged in a mobile telephone conversation. We are freed, to a certain degree, from the normal conventions regarding copresent interaction. There is both the engagement with the telephonic partner and a signaling to others that we are occupied. Thus, ducking our head down and perhaps odd space juxtapositions, such as standing in a corner of the room, are signs to those who are physically copresent that we are not to be disturbed. In the same way that a couple's walking hand in hand is a sign to others that they have a certain type of status, the closed posture of the mobile telephone user is a quasi-necessary indication to the immediate milieu that the individual is not available at the moment for normal interaction.

Gaze is a particularly strong element when considering the willingness of an individual to participate in a social interaction. In this case, it is another strategy for managing our aloofness (Murtagh 2002). The tendency for those who are using a mobile telephone in public to avoid looking into the eyes of other persons is quite strong. Indeed, as part of the data collection for this analysis I would try to catch the gaze of several dozen mobile telephone users while walking down the street. Generally this was not possible. If I did catch the eye of a mobile telephone user, her or his glance was only a very quick "navigation interaction."

In addition to gaze, facial gestures such as smiling and looks of concern can be seen as a way to mark our removal from the social situation. Those using the mobile telephone often smiled, knitted their eyebrows, and used other facial gestures that seemed to be superfluous given the mediated form of interaction. Clearly, many of these outward signs are a reflection of our inner joy or concern vis-à-vis the content of the conversation. As with the use of gaze, they are also signs, perhaps unintended, of the engagement of the individual in the mobile telephone conversation. Thus, they further underscore the social distance of the individual from the physical location.

The uttering of a half sentence in the observation cited earlier is also of note. It seemed to indicate that the call was somewhat expected and that it played on an understanding of the two copresent partners that something like this might happen. It underscored the transition of the man's attention from the common task of searching for a book with his partner to the telephone conversation that had already been announced by both the audible ringing and his search for the device in his pocket.

The man further partitioned the audible portions of the telephonic interaction by wandering off to other, less trafficked parts of the store.[7] This points to another issue that is often commented upon, i.e., the colonization of the public sphere by mobile telephone users. The use of space, and behavior within the space, mark our role in the situation (Hall 1973). A person who uses a mobile telephone in public areas often moves into less trafficked parts of the space. In order to examine the degree to which this is an important strategy, I tested the degree to which a person was willing to protect his or her "territory" by moving into the immediate area of the telephonist. I did not, for example, march up in front of the individual and stare into his or her eyes. Rather, 10–12 times in stores I would start to examine the wares immediately beside the telephonist. The result was always that the individual yielded the space to me by wandering off to another, less populated area. If I were to repeat the experiment with the same subject, the person would again wander off. Aside from slightly bothering the subjects of the trial, this points to the awareness that the individuals seemed to have regarding their surroundings and the importance of maintaining a buffer around the self. Even though they were involved in their telephone conversation, they were, at some level, still aware of the things going on around them. In addition, they were willing to surrender their space and move on. Unlike the metaphor of the boorish egomaniac loudly talking on his phone, this experiment seems to point out that mobile telephone users are alert to the need to maintain the space around them if possible. In addition it points out that the barrier is not absolute.

The point here is that all of this interaction has been arranged within a relatively quick transition period as we move from our physical surroundings into the more virtual world of a telephone conversation. It demands that the telephonist, the accredited members of his or her colocated party, and, at a more diluted level, others who are nearby readjust themselves to the new situation. It is clear that this is not always easily achieved.

In the previous observation the telephonist could retreat from the immediate situation. If this is not possible — when, for example, sitting on a bus — a subset of these strategies, such as averting the gaze and a closed posture, can be used.

> **Observation:** Three people were sitting in a bus. A woman was sitting next to a window on the right side of the bus. To her left sat a man in the aisle seat, and a second man sat across the aisle, again to the left. The three got on at the same stop and continued the conversation from when they were standing waiting for the bus. The woman's phone rang and she reached for it immediately. As she dug for the phone the two men turned toward each other and started talking to each other. The man in the middle turned his head and shoulders toward the man sitting across the aisle, further marking the temporary exclusion of the woman.
>
> The telephone connection somehow failed before the woman could start the greeting sequence (perhaps she accidentally touched a button when digging for the phone in her bag.) Thus, she again became available for inclusion in the threesome. After about 20–30 seconds her phone rang again. She muttered a small "oh!" as she again recovered the phone. The two men again closed her out of the circle, both in terms of the direction of their comments and their body language. She answered the call and turned toward the window of the bus during the actual call.

As with the previous observation we see here the nimble staging of the telephonic event. As it turns out, the woman and her friends needed to go through the process of readjustment two times before the actual telephone call was answered. Unlike the previous example, there was no opportunity for the woman to physically distance herself, and so both she and her friends used body language to provide each other the space needed for the call.

Management of the Local Situation During the Call

The barriers erected in the case of mobile telephone calls are a tacit recognition of the need to manage the situation. Various interkinesic and interproximic devices are erected as barriers between the telephonist and his or her social circle. While this is an important strategy in many social situations, it is also clear that the barrier is not absolute. During the period of the call, accredited copresent partners often assume a suspended status. They are not dismissed; rather, they are left hanging. Only after the conclusion of the call can they resume their earlier status. However, in certain situations they can intrude into social interaction.

> **Observation:** A pair of teenage girls entered a streetcar and sat together. They discussed things until one of their mobile telephones rang. That girl took it out and started a conversation with another person. She was quite animated in the conversation and used several gestures, including hitting herself in the forehead with the palm of her hand when apparently reminded of a forgotten issue. The second copresent girl withdrew and took on a "streetcar" face. The conversation continued past two or three stops, and eventually the second girl rose to get off. The girl on the phone remained seated. The two copresent girls went through a quasi-pantomime of farewell gestures, including a hug and touching cheeks, along with the accompanying "kiss" sound as used in a greeting in France. All the while, the girl on the mobile telephone maintained her part of the conversation. One stop later, the remaining girl completed her call, replaced the mobile telephone in her bag, and got off.

In this case, the girl who received the telephone call had shifted her attention to the phone call and had temporarily parked the copresent friend. However, when the time for parting arose, she was able to manage both the telephonic and the copresent social interaction. Indeed the teen telephonist in this observation was quite agile in her stage management. In this case, the interruption was only a short interlude in the longer telephone conversation. In some situations the copresent interaction can even overtake the telephonic.

> **Observation:** While out walking in a popular forest area near Oslo (Sognsvann), a man pushing a baby carriage was overtaken by another party. One woman in the second party was talking on her mobile telephone. The man pushing the baby carriage and the woman on the mobile telephone recognized each other. The woman on the phone moved the microphone away from her mouth and said ironically to the man pushing the carriage, "This is my image." The man greeted her, and she lowered the mobile telephone from her ear to about shoulder height, indicating that she was caught, in a sense, between the telephone and the colocated interactions. The woman on the phone said that her telephonic interlocutor was "Heidi." This indicated that it was a common friend. The man thought that it was the "Heidi" who lived in one location, but after several interactions it became apparent "Heidi" had moved. All the while Heidi was on the line, presumably hearing bits of the conversation since the mobile telephone was still being held at about shoulder level away from the mouth of the woman. Eventually the woman informed Heidi of her chance meeting with the man pushing the baby carriage. The man and the woman continued their conversation, and there were occasional attempts to bring Heidi into the circle, but these were more and more strained. Eventually the woman asked Heidi if she could call her back later.

This, along with the material from my "invasion experiments," shows that while the ability to immerse oneself in a mobile telephone conversation is quite strong, it is not impermeable. Nonetheless the status of copresent individuals is a problematic issue. Their presence and the degree of their involvement need to be arranged and managed. This is difficult since the behavior appropriate in the one situation is not appropriate in the other. The different relationships we have with copresent partners and our telephone partner mean that the negotiation of topics, the depth, passion, and emotion with which we can address the one party, and the range of common understandings will probably not be the same for the other (Garfinkle 1967). There is a certain dissonance when judged from the perspective of classical notions of talk, dialogue, and narration that present a single perspective to the audience.

When considering copresent individuals who are accredited members of the telephonist's circle, there are several strategies available. They can be "parked," in that they are asked to engage themselves in some type of waiting strategy and assume a form of studied nonobservance that is easily discarded when the other summons them back (Schwartz 1977). Here the companion is left to maintain his or her own illusion of indifference or studied noninvolvement. This was the initial status of the copresent third person on the streetcar in the observation just cited. The telephonist directed her attention away from the third person and did not announce her presence to her telephonic partner. It may be necessary for the telephonist to grant a temporary backstage status to those who are copresent, and perhaps take them into his or her confidence in various conspiratorial ways (Goffman 1971, pp. 220–222). Thus, they will be allowed insight into the staging of the telephone conversation and may even be brought into collusion with reference to its staging and the telephonists "real" feelings toward the telephonic partner through the use of winks, rolling of the eyes, and the like. In the short term this allows the completion of the telephone call; however, once belief in the first performance is suspended, rebuilding its façade may be difficult, particularly if the original audience is only grudgingly willing to accept its backstage status.

Another alternative is that the presence of the third person is announced but that the third person is not invited and does not take the initiative to participate in the telephone conversation. This may indeed be a tacit way to limit the type of themes that may be taken up, the depth with which they will be examined, and the tenor of the conversation.

It is also possible for copresent individuals to be included in the conversation. The third person may even make open side comments (Lohan 1997). The others may feel free to come up with interjections, stage whispers, and disruptions (as in the case of small children that need assistance) and can also key on the flow of the conversation. While they are not a central part of the telephonic interaction, they can read the gestures and the gaze of those who are on the phone to determine the appropriate degree of their own inclusion.

Finally, the third person can completely overtake the situation. In the case cited earlier the telephonic partner was cut off by the appearance of a copresent friend. Others have reported the opposite, in that the third person actually took over the role of being the telephonist (Weilenmann and Larsson 2002). In each of these cases, however, there is the need to manage and adjust the staging of the interaction so as to provide the various parties the appropriate level of insight into the proceedings. This can be awkward given the complexities of the interaction.

Reemergence into the Local Setting

Eventually the telephone conversation approaches its completion. Just as the ringing of the telephone signals a transition from one social situation to another, the completion of the call also heralds another transition. As discussed earlier, the parting sequence in the telephone contains several parts. In essence the telephonic partners negotiate the end of the interaction. They find that they have completely covered the various themes, and eventually one initiates a farewell sequence. In this interaction, one may state an open sentence, such as "OK" or "Well" followed by a pause. If the other individual does not fill in the pause with a new theme, he or she may countersign with a similar phrase. In essence, this is a negotiation to begin farewells. Following this they may make a quick summary or add some final comments. This done, they will exchange farewells and hang up. Thus, there is not a simple, quick closing of the conversation, but a social process through which to go in order to be seen as a competent conversationalist.

The interesting thing from the point of view of mobile telephony is that the closing sequence can also inform the copresent individuals that they must drop the façades they had assumed upon the initiation of the mobile telephone call and prepare themselves for further interaction with the telephonist. We can see these elements in the following observation.

Observation: Two men were sitting in an airport waiting area toward the end of a work day. They appeared to be work colleagues en route home after a business trip. One received a call and answered it. While he was talking, the other man occasionally looked on but also freely glanced about at other passengers, at the screen display showing flight departures, and at the tabletop in front of him. When the man on the phone started the farewell sequence, the other man looked toward the first. After saying his farewell, the telephonist pushed the "hang up" button on the phone, took a quick glance at the display, and lowered his phone (still holding it in his hand). The telephonist glanced back at the other, smiled, and seemed to give a short summary of the call. The two men fell back into discussion.

Here it was obvious that the other man was listening enough to know that the farewell sequence had begun and that he would soon be back on stage. After the end of the conversation, the telephonist seemed prompted to provide the other with a short summary of the interaction, perhaps as a partial legitimation of the need to suspend the conversation. This is a type of repair work that is used to bridge the transition from the telephonic to the copresent interaction.

Thus, we have gone through the cycle of the telephone call. We have moved from the initial ringing and the perhaps hectic rearrangement of the social order, through the management of the various publics during the body of the interaction, and finally to the resolution of the call and the reemergence of the individual into the local setting. In each phase, the introduction of the mobile telephone raises various issues that need to be resolved. In each case the novelty of the situation means we are often improvising these solutions on the spot. Thus, they can have a stilted and awkward feeling.

Forced Eavesdropping and Being Embarrassed for Others

The accessibility to others' telephone conversations exposes us to what might be called *forced eavesdropping*. Informants often felt that they would hear too much, a sort of coerced eavesdropping. The need to guard against this was seen as a problem. This is often considered a breach of decorum, in that we are being exposed to the personal situation of nonintimates.

Several focus group informants of the respondents noted that the audience to a mobile phone conversation is not only our immediate partners, but also those who are within hearing distance of the conversation. Often, access to information is little more than superfluous, as in the case reported by one informant who said, "[The user of a mobile phone] yaks away about trivial matters." In other cases, there may be the fear that the conversation will turn to risky issues that may not necessarily be intended for others. Thus, the audience is provided involuntary access to portions of the phone user's life, a clear violation of Gullestad's managed unreachability (1984). In the words of one respondent:

> *I think it is fine that the telephone is stationary at home. A conversation should be between two persons. I think it is unfortunate that others are there. (Grethe, 63)*

Informants indicated that forced eavesdropping prompted what some called an "embarrassment for others." This points to the idea that embarrassment is not simply a psychological condition that might, for example, be the result of weak ego development. Rather, it suggests that we can use embarrassment as a way to examine the social situation.

Goffman suggests that there are two types of embarrassment (1967). The first type is seen in longer-term states of affairs in which we do not have a legitimate sense of our role in the situation. Thus, a freshman undergraduate at a faculty party would not necessarily have the aplomb to handle the situation. The second type of embarrassment is the relatively rapid realization that we are caught out of character. The classic example for men is realizing that they have been seen in public with their fly down. Here we go from a state of believing that the situation is under control to being unsure of who has seen us, the degree of the gaff, and the extensiveness of the repair work needed. It is this type of situation that is of most interest here.

In the context of eavesdropping, hearing the conversation of others, or being in the position to hear the conversation of others, is a tricky social situation. While eavesdropping is not the worst social faux pas, it is nonetheless awkward. This is potentially embarrassing to the person doing the eavesdropping, the persons who expose them, and the person(s) who have been overheard. However, there are several variations to be considered here. If there is a willful element in the eavesdropping, i.e., if you are spying, then this is an openly malicious motivation. However, if you only appear to be eavesdropping but are actually carrying on with other things, then there is a disjunction between the appearance and the intention of the eavesdropper. Nonetheless, playing on W. I. Thomas's dictum, if eavesdropping appears to be real, then the consequent indignation of the overheard person can truly be real. Thus, as we move through society we need to protect ourselves against these types of potential claims against our propriety.

The third variation, i.e., forced eavesdropping, is more to the point with regard to mobile telephony. In this case, the potential eavesdropper is, in a sense, overrun by the person being eavesdropped on. The physical situation or the level of conversation is such that the noncredentialed person cannot legitimately hold himself or herself outside of the interaction. This happens, for example, when a microphone is left on after a presentation or performance. As the presenters shift back into their roles as normal individuals and perhaps give their unvarnished impressions of the performance or of the audience itself, the forced eavesdropper has no way to withdraw from the situation.

In the first two forms of eavesdropping, it is the eavesdropper who is particularly exposed to the potential for embarrassment. He or she has been caught, and thus his or her composure is at risk. In the third situation, i.e., enforced eavesdropping, the people within the conversation may be embarrassed, but perhaps the more interesting thing is that the eavesdropper(s) are also embarrassed. This is a special type of embarrassment, namely that we are "embarrassed for others" that is, we are embarrassed for the sake of those persons who are forcing us to be eavesdroppers.

There are several points here. First, while the cause of the embarrassment is the actions or the situation of an individual, we are also embarrassed for the whole situation. Regardless of how loosely defined the situation is, be it a group of people waiting for a subway or a group attending the gala dinner for this year's Nobel Prize recipients, each individual has a responsibility for maintaining sociability in the entire situation. When an untoward event arises, the individual may feel, more or less strongly, an angst that the whole situation is at risk. A second point is that the potentially embarrassed person has not yet actually been caught. Rather, the person who is out of character has not yet recognized his or her faux pas and thus has not yet lost face. This is a type of pending event that the observer feels will become embarrassing as his or her situation becomes obvious. Clearly there is an implicit social tension here.

Beyond the spectacle of the impending situation, we can pose the question of why the viewer, or the forced eavesdropper, feels embarrassment? After all, it is not that person's reputation that is at stake. I suggest that we become embarrassed for another because the noncredentialed person who is being forced to eavesdrop has seen someone in a position in which that person may have been or potentially will be.[8] Building on Mead and Cooly, Giddens suggests that identity is a reflexive project. Embarrassment then is when there is a schism between our imagined identity and a disjointed identity made obvious to us by some faux pas, either intended or accidental. Following from Giddens, embarrassment for others is a vicarious version of the same thing. It is a threat to our identity that is linked to our identification with the soon-to-be embarrassed person.

One final point here is that the embarrassment felt by the observer and the "actual" embarrassment of the person being caught need not correspond to each other. Even if the person who is being caught is made aware of his or her predicament, it may not have the same meaning as for the person viewing it. While the viewer may cringe at the behavior of the party animal openly flirting with a happily married woman, this need not be the same experience for the two individuals directly involved. That is for them to decide. The observer is free to be as embarrassed as he or she wishes. Indeed, the engaging nature of a telephone conversation may make the telephonist oblivious to the broader social situation and to the eventual need to be embarrassed. The phoner may never become embarrassed.

To the degree that overhearing a mobile telephone conversation exposes us to forced eavesdropping, there is the potential to create the discomfort of embarrassment for the involuntary audience. It seems there is a sense in which the public use of the mobile telephone will threaten our identity.

Conclusion

Mobile telephony has the ability to disrupt the structure of social interaction at several levels. At the broadest level, the device challenges the decorum of established social settings, such as those in restaurants. At a more microsocial level, there is a range of disturbing elements in the way that we manage interpersonal interaction vis-à-vis the mobile telephone. Finally, for the individual actor in a social situation there is turbulence caused by issues such as forced eavesdropping. At each level various issues associated with the use of the device were elaborated.

After this examination, however, I am prompted to ask if all of this is really a problem? The answer seems to be both yes and no. To the degree that we have not collectively developed routines and rules for making the transition from colocated to virtual social interaction on short notice, it is difficult. To the degree that we feel embarrassed by the forced snooping on other's conversations, the public use of the mobile telephone is a problem.

This said, we are remarkably flexible in our ability — eventually — to accommodate technologies. Standage (1998), Marvin (1988), and Fischer (1992) provide insight into the social disruption accompanying, respectively, the telegraph, electricity, and the telephone. With time, the telegraph has been forgotten and the two latter innovations have become taken for granted. The ringing of a home telephone is handled with well-oiled routines — if not excitement in the case of lovelorn teens. Switching on a light is only occasionally the focus of social ritual, as, for example, when lighting the family Christmas tree or dimming the lights for a "special" dinner. The mobile telephone may, in all likelihood, go the same way.

7

CHAPTER

Texting and the Growth of Asynchronous Discourse

Introduction

Since the late 1990s, the use of texting via mobile telephones has seen phenomenal growth. Like the legendary rabbits introduced into Australia, text messaging seems to have exploded into an empty niche. In countries using the GSM standard, it has been via the use of the Short Message System (SMS). There is the equally prodigious use of the text message function in the DoCoMo I-mode and other mobile telephony standards. Indeed, text messages have become central to teens' use of mobile telephony. Among some groups, texting is the alpha and omega of mobile communication.[1] According to one teenage focus group informant, "Mobile [telephoning] equals SMS for me, nothing more." (Dagfinn, 17)

At the time of this writing, statistics for Norway show that on average more than 335,000 messages are sent every hour and more that 8 million every day, and this in a country with only about 4.5 million inhabitants. If the distribution were even across the whole population, every Norwegian would send about two SMS messages a day. As it is, the average mobile *user* in Norway sends almost 70 SMS messages every month (PT 2003). The usage rate is among the highest in the world (Nurmela 2003). However, there is also high usage in other countries. In what has grown to be a new tradition, about 2.5 messages per capita were sent in Italy to celebrate New Year's Day 2003. On a worldwide basis the number of text messages went from 4 billion in January 2000 to 20 billion in June 2001. According to *Cellular News* there were 95 billion SMS messages sent in the fourth quarter of 2002 and more than 366 billion sent during the whole year. If the use of texting were evenly divided among the world's population, every sixth person would have had to send one message a day to reach this sum. Interestingly, the United States lags behind (Riedman 2002). To use the same measure, the traffic in the United States in 2002 could have been generated by a daily message sent by 9 persons in 100.[2]

The breadth of "texting" includes more than the simple transmission of text-based material from one mobile telephone to another. Increasingly, "text" messages are including photographs, sound files, and other attachments. In addition, text messages are being adapted into various forms of gaming and TV voting, and they are starting to move across platforms. Both I-mode and enhanced GSM telephones can send messages to Internet addresses, and vice versa. We can also engage in various types of "chat" and instant messaging sessions from mobile telephone terminals, albeit at a lower transmission rate.

Obviously, even within countries, use is not evenly distributed. Some groups are heavy users; others practice total abstinence. Norwegian teens are among the heavy users (Figure 7.1). It is the preferred form of mediated interaction, surpassing instant messaging, e-mail, voice mobile telephony, and even traditional landline

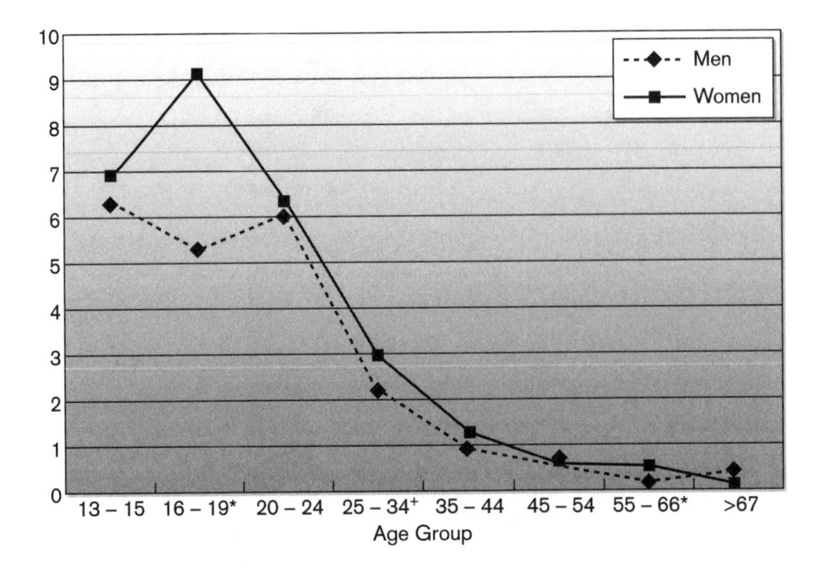

FIGURE 7.1 Mean number of SMS messages sent per day for SMS users, by age group and gender, Norway, 2002 (* = Sig. < 0.05, + = Sig. < 0.10).

telephone calls. Teens are also a major user group in Italy, Finland, Japan, Korea, and the Philippines (Paragas 2000; Ito 2003; Hashimoto 2002; Rautiainen and Kasesniemi 2000; Mante-Meijer *et al.* 2001). These statistics bespeak an impressive form of communication.

Given this popularity, we have to wonder what is so beguiling about text messages. Much of the answer is that text messages are relatively cheap and convenient. Text messaging allows us to maintain contact with friends and colleagues, but at the same time it is inconspicuous. Texting allows us to be expressive even in situations where other forms of communication are not appropriate. For example, we can text when sitting on the bus, in the classroom, or, in the case of socially starved teens, under the covers late at night. It allows us to coordinate everyday activities, to send endearments, get quick answers to questions, and keep one another up to date concerning the large and small events in our lives. In Europe it is often seen as the preferred medium for the maintenance of friendship networks (Smoreda and Thomas 2001a; Crabtree *et al.* 2002). It is used to fill up the odd free moments of the day. In the words of a 17-year-old boy, "Often when we are sitting on the bus or subway it is boring, and so we can write messages and that entertains us in those boring moments."

Beyond the normal maintenance of social groups, texting has been credited with assisting in the coordination of social movements, such as the revolt against Philippine President Joseph Estrada, bicycle activists in Boston, antinuclear campaigners in Germany, and the organization of the antiglobalization "battle in Seattle." (Plant; Rheingold 2002).

When designed in the 1980s, the system drew on the inspiration of the already existing paging services. Two of the major technical innovations were the ability to send as well as receive messages and the "push" function; i.e., beyond turning on our mobile phone, we do not have to go through a log-in sequence before receiving messages. The people working on the specification of SMS drew on the image of taxi and parcel delivery fleet management in the United States. Beyond the traditional dispatch function, they conceived of a type of peer-to-peer interaction (Trosby 2003; Trosby).

In some ways, text messages are an odd construction. They are difficult to write, since there is no traditional keyboard or writing instrument. The message is limited to only 160 characters; the displays for reading the messages are small. Finally, transmission relies on terminals limited by poor batteries. Nonetheless, many of us — most particularly younger users — have adapted. We have pared down our messages into a cramped telegraphic style that may be more linguistically akin to speech than to writing.

In general, it seems that text messages focus on who, what, where, and when. There is perhaps less of a focus on how, why, and, eventually, how we feel. Drawing on similar research done on e-mail (Baron 2000, 2001, 1998; Yates 1996; see also Herring 1996), the material examined here shows that in spite of the fact that men were early adopters of mobile telephones (Ling 2000c), it is among teen women that we find the literary stylists of texting culture. In the words of a focus group informant, "Most of the messages I get from boys are pretty short because they don't think it is so fun to sit there and punch in on the phone. That is more of a girl thing" (Erin 17). It seems that her comments are on target. The material here shows that teen and young adult women write longer messages that are more complex. They are more likely to include literary flourishes, such as in capitalization, punctuation, and emotional elements, and they are more inclined to include refined formalities, such as salutations and closings, in their text messages.

Texting has a varying role when it comes to the other general themes of mobile telephony, i.e., safety, coordination, access, and the disturbing influence. In general, texting is a coordination tool more discreet than other forms of interaction. In sum, texting is a living medium. It has sprung onto the scene in many countries, and, in one form or another, asynchronous mobile text communication seems here to stay.

The locus of texting is among teens and young adults, in particular among women users. Using data from the United Kingdom, Italy, Germany, and Israel, we consistently find that that younger users report sending more text messages than other groups.[3] Data from Norway in 2002 also shows that women[4] and teens/young adults[5] are the most enthusiastic users of SMS. In 2002, more than 85% of teens and young adults reported sending SMS messages on a daily basis.[6] By contrast only 2.5% of those over 67 reported using SMS with this frequency. While 36% of the men reported daily use, more than 40% of the women said that they send SMS messages on a daily basis. The same general trend is reported in Finland, where women are also reported to be heavier users of SMS than men (Rautiainen and Kasesniemi 2000).

The data also indicates that the same groups were the most intense users. Women[7] and teens/young adults[8] report sending significantly more messages, on a daily basis, than their counterparts. As we can see in Figure 7.1, women 16–19 years old who were SMS users — and only about 2% of this group does not use SMS — reported sending a mean of slightly more than nine messages per day. The material from the *Ung i Norge* (Young in Norway) study shown in Figure 7.2 also confirms that teen women — in particular younger teen women — are more active on the SMS

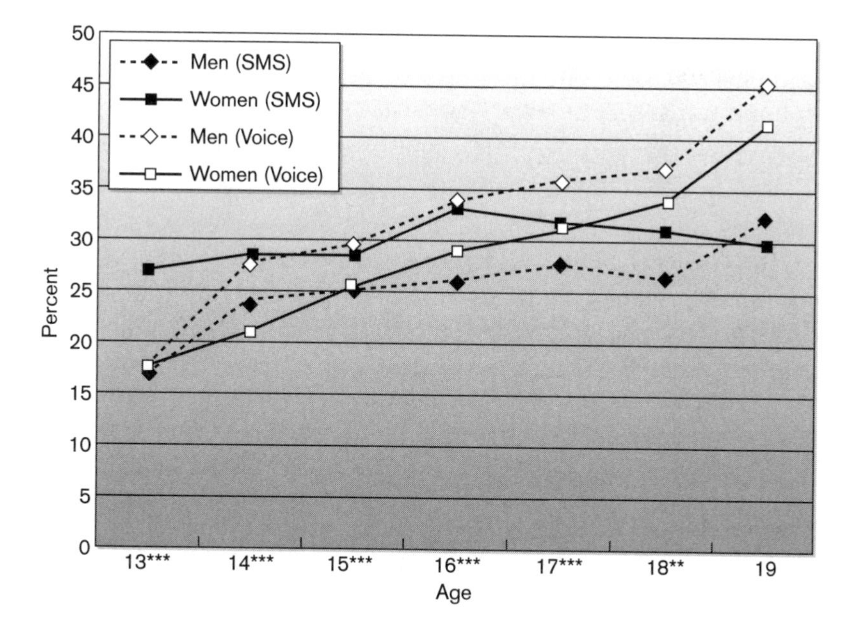

FIGURE 7.2 Percent of teens reporting sending more than five SMS messages and making more than five mobile voice calls per day, by age and gender, Norway 2002 (*** = Sig. < 0.001, ** = Sig. < 0.01).

front than are teen men.[9] There is a strikingly similar profile if we examine the use of messaging services in Japan. The peak in use is among women 20–24 years old. More than 90% of these young adult women reported using text messages, whereas only about 10% of women 50–59 years old reported this (Hashimoto 2002). Thus, if we compare the usage rate of young adult women to that of their mothers or grandmothers, chances are good that the older guard does do not use SMS; and if they do, their usage rates are literally an order of magnitude below that of their progeny.

The Growth of Texting

Texting and the Individual

Why is SMS so popular? This question can be addressed from the standpoint of both the individual and the group. From the perspective of the individual user, texting has several advantages. It is relatively inexpensive, convenient, and unobtrusive and can be used for strategic purposes in some situations.

The most commonly noted advantage is that texting is relatively inexpensive and easy to budget for. Teens who generally must pay for some or all of their own mobile telephone use are quite aware of costs. Indeed, this frugality spawned the rise of SMS in Norway.[10] We can see this attitude in the comments of Nora (18): "A little message is a lot cheaper to send," and of Rita (18): "It is so expensive to call with a mobile telephone; it is not worth doing it." Data from Germany, the Netherlands, the United Kingdom, Japan, and Norway all point in the same direction, namely, that it is cheaper to send a text message (Skog and Jamtøy 2002; Smoreda and Thomas 2001b; Taylor and Harper 2001; Ito 2003; Doering 2002; Grinter and Eldridge 2001; Hashimoto 2002; Ling 2000c).

From the perspective of the user, a voice telephone call is difficult to price. The length of the call is variable, since we go through a more or less elaborate greeting, the discussion of various topics, and a ritualized farewell (Schegloff and Saks 1973; Saks *et al.* 1974; Veach 1981). All of this takes time. In addition, tariff rates are often complex, and we are unsure about which tariff is applicable at what time and for which type of call. There is an uncertainty and indeed often a tendency to overestimate of the actual price of voice telephony (Ling and Hareland 1997). Thus, from a budgeting perspective, there is a real advantage to text messages, in that they are a concise event and are unit priced.

Texting is seen as quicker and more convenient than voice telephony (Grinter and Eldridge 2001; Smoreda and Thomas 2001b). Although it is suspected that sending a text message takes a long time, many users report that this is not the case. Accomplished texters can theoretically key in messages at speeds approaching 30–40 words a minute, particularly if they use the predictive typing programs included in many mobile phones (Silfverberg *et al.* 2002).[11] Since, as we will see, the average text message contains about six words, this does not represent a significant investment of time. Further, you can send a message at your convenience. Sending a message does not assume that you are coming into direct contact with the receiver; rather, it is often the case that you are simply "posting" a message for later scrutiny. Finally, the messages are rarely pretentious. As we will see later, the "interaction" is shorn of many of the social fine points encountered in verbal interactions. As already noted, texting allows you to fill in odd moments with social interaction. Thus, while you wait for a bus or are sitting in a streetcar you can engage in social interaction and fill that time. You can chat or joke with remote friends, gather news, or coordinate further interaction in these otherwise unoccupied spaces in our life.

Texting is relatively inconspicuous in comparison to voice telephony (Ito 2003; Haddon 2000). We can text others as a type of nonintrusive or even an illicit secondary activity in school or late at night, when we should be sleeping (Taylor and

Harper 2001). If, for example, a teen turns off the ringing sound on his or her mobile telephone, nobody is the wiser that he or she is sending or receiving communiations. Analysis I have done shows that teens send and receive text messages in class at school and throughout the night. Indeed, data gathered in Norway in 1999 showed that 24% of the teens included in the study used their mobile telephone during class, despite official restrictions. In another study from 2002, we found that around 20% of teens say they send and receive SMS messages after midnight on a weekly basis.[12] Thus, it seems that today's youth have replaced the flashlight with a mobile telephone when hiding under the covers. Instead of solitary reading, they are engaged in social networking.

Texting, and indeed mobile voice telephony, is more individualized than traditional telephony. With a traditional telephone, we call to a home or other location and then must often request a conversation with a particular individual. The mobile telephone, however, is highly personalized. When we send a text message to a certain telephone number, we expect that it will reach a specific person regardless of where the person is at that moment. Mobile telephony and texting have allowed us to work these interactions more carefully into our lives.

Text messages are asynchronous; that is, the sender does not need to engage the complete attention of the receiver in order to communicate. In addition, the sender can compose, edit, and send a message. The receiver can pay attention to it as time allows. This dimension of texting has also sparked a type of quasi-mediated form for developing romantic relationships (Ling and Yttri 2002; Grinter and Eldridge 2001; Ling 2000a). In this process, initial contact is made face to face at, for example, a party or other social gathering. As a part of the contact, you exchange mobile telephone numbers. In the days that follow, you send a text message to the other, reopening the contact and further developing the potential relationship. The contact usually takes the form of a noncommittal question, since this implies a response. Here the use of texting allows one to compose and edit our message, perhaps with the aid of friends. In Goffmanian terms, the indirect nature of text messaging gives the teen a chance to arrange "face." Rather than fluster through an awkward telephone conversation, the teen can carefully edit the message before sending it. We see this in the comments of Ida.

> ***Ida (18):*** *Then you do not have to use your voice, which can shout or break up. You have to have time to think.... You always use it in situations like this because it gives the other person the chance to think through and answer no. If the person is on the phone, it not always so easy to answer no. (Emphasis added.)*

The message is sent directly to the focus of our interest. Since the message is not a physical object, such as a note, you need not meet the other in order to deliver the text message, nor do you need to engage an intermediary to play Cupid. You need not call the home of the person and perhaps meet skeptical parents before being granted permission to speak to the potential new paramour. If the "other" responds positively, you two can further develop your interest via texting before, perhaps, going over to more synchronous interaction.

Thus, the ability to think through the messages carefully and to manipulate the delays in the interaction mean that text messages are well suited to the early phases of relationships, when the main focus is on exploration of common interests (Berger and Kellner 1964; Parks and Roberts 1998). They allow for direct access, but avoid the problems of being on the open stage. You can compose specific messages for the object of your interest and, at the same time, avoid giving the wrong signals. As the relationship moves into other phases, such as that of building trust, more synchronous communication is often preferred (Ling 2000a). This technique has some kinship to the development of online relationships (Parks 1996; Smoreda and Thomas 2001a; Baym 2002; Parks and Roberts 1998; Lea and Spears 1995; Walther 1993), the difference being that it employs both mediated and face-to-face interaction. Thus, the textcentric dimension may mean a different degree of engagement in the communication and less of a focus on physical appearance.

Interestingly, text messages are also used at the end of a relationship, albeit not without morally imbued comment. It is common to hear of individuals who receive the final "Dear John" announcement via text messages (Taylor and Harper forthcoming). There is a certain symmetry here, in that the texting serves as a type of barrier both at the outset and at the resolution of a relationship. Where at the outset of a relationship texting serves to restrain the tempo as we carefully think through the messages we send, the use of texting at the end of a relationship, in effect, reerects the barriers.

Texting and the Group

Beyond the advantages for the individual user, texting can be seen from the perspective of the group. Indeed, texting is on the face of it a form of interaction between individuals. Thus, as we have seen, the messages help to coordinate, inform, and generally care for our social contacts. We arrange parties, send and receive various forms of personal news, provide compliments and endearments, and the like. In this respect, texting has become a real link in the social network, a way to maintain a "background awareness" as to what is happening within our social sphere (Ito 2003).[13]

Regarding the use of the mobile telephone for colocated individuals, the mobile telephone often plays a role in local social interaction. Various authors have reported that composing and reading the messages can be a collective affair (Kasesniemi and Rautiainen 2002; Taylor and Harper forthcoming; Weilenmann and Larsson 2002); indeed, the telephone itself becomes a significant object in the social interchange of the copresent group (Taylor 2003). Thus, the production of SMS and the use of the telephone can be seen as a collective activity of copresent individuals.

Further, text messages, in particular the "grooming" messages described later, can be seen as a type of gift giving (Taylor and Harper 2001; Ling and Yttri 2002; Bakken 2002; Johnsen 2000; Taylor and Harper forthcoming). From this perspective, they help to crystallize the form of the relationship, and they serve to set a special frame around the situation in which the exchange takes place. They can mark what are literally once-in-a-lifetime events, as in this message sent by a 41-year-old woman: "*Congratulations with the birth of your boy.*" A text message can be a remembrance associated with a specific occasion, as in this SMS sent by a 31-year-old male: "*Thanks for the party last weekend it was a success.*" Or it can be small endearment used to mark an important relationship, as this message sent by a 17-year-old woman, presumably to her boyfriend: "*Sleep well sweetie. Call me tomorrow when you get up. Love you.*"

In each case, the message is a part of a broader ongoing relationship between the sender and the receiver. None of the messages is focused on the planning of future interactions in any direct way. Rather they are focused on the maintenance of the relationship in the form of small ritual interactions. The exchanges in themselves are specific events that summarize, in miniature, the relationship between sender and receiver — that is, the giver and the receiver of the gift. In this way, they contribute to the broader relationship, in that they enmesh the sender and the receiver in a situation of reciprocity. Failure to reciprocate, and failure to do so in a timely manner with a message of similar value, can indicate a breach in the relationship (Taylor and Harper 2001). If you wait too long to reply, if you send a short unimaginative response, or if you send a response not written especially for that occasion, such as a canned joke, then you are on thin ice in the reciprocity department.

Another way in which texting is an integrative activity for the group is that it is in many respects a common cultural artifact of the current generation. Indeed teens and young adults have made texting a part of the experience of youth. Teens have established rules as to what can be said (prank messages vs. other types of more serious messages), when it is acceptable to send messages, the form of the content, and the staging of texting interactions. There are rules as to what can be said and communicated in a text message, when to call, and when to reply. There is a sense as to the appropriate cadence of the interaction and what pauses are acceptable.

In sum, there is a loosely defined ethic associated with the use of text messages. It is a living feature of teens' everyday life, it is the expected norm within the group, and it is a symbolic activity that in itself defines the teens vis-à-vis each other and older generations. The use of the system in diverse, unorthodox situations can also produce a certain esprit de guerre. Illicit texting in school, late at night, or in other, more staid situations can provide the teen with a sense of participation in a larger social movement. In these ways, it is an integrative activity for teens.[14]

Obviously, there is a dark side to the use of texting. At the innocent end of the scale, texting in school can divert students' attention away from the academic issues at hand to the organization of social events and the like. Its use in the evening can steal from needed sleep. The complex planning of social events can be manipulated in order to exclude teens from various social events, and the use of the mobile phone can play into the power relations between parent and child (Ling and Yttri 2003).

Texting has also been used to bully others. This plays on the extremely personal nature of the medium and on the fact that messages can be sent anonymously. Texting has also been used to facilitate various types of illegal and quasi-legal behavior, including prostitution, smuggling, broadcasting warnings as to where speed traps are located, cheating on exams, and spreading the word on "home alone" parties that turn into ransackings.[15]

What Is Being Said, Who Is Saying It, and How They Say It

Content of the Messages

Beyond the sheer numbers of messages and their functions within the group, we can ask about what is being said in the messages. Much has been made of dating, romance, the organization of protests, and other stirring forms of behavior. In spite of these rousing images, many analyses find that the content of messages is generally focused on everyday tasks.

SMS has found a functional niche in our communication needs. In the words of Gro, an 18-year-old woman, "I use it if I am just going to send a short message, for example, if I am just going to ask if they are going to go out. It goes a lot faster." She continues by saying, "I send messages if I am planning something, if I am bored, or if it is something that is important." Her reported use spans several of the categories found in the data, i.e., coordination, planning, as a type of entertainment, and information collection. Data from Germany, the United Kingdom, and Norway shows that these themes are common (Grinter and Eldridge 2001; Skog and Jamtøy 2002; MORI 2000; Taylor and Harper 2001; Doering 2002).

To get a better grip on the actual use of texting, it is worth looking into the content of the messages. To this end, Telenor gathered a corpus of 882 messages from a random sample of Norwegian SMS users.[16] The material here shows that various types of coordination, "grooming," and questions/answers make up the bulk of the traffic. The material examined here indicates that about one in three text messages was used to coordinate various meetings and activities. Another 17% of the messages were "grooming" messages, or statements in which there was no real planning or instrumental information exchange, but rather a simple — usually positive — remembrance. Finally, various questions and answers made up about a quarter of all messages in the corpus.[17]

We find sociodemographical differences. Using at a slightly more refined categorization, we find that men are slightly more prone to using short, one-word answers in their SMS messages.[18] When it comes to using SMS messages to plan activities, men are more likely to use them for planning activities in the middle future,[19] as are older teens and young adults.[20] Women, however, are more likely to use SMS to make plans for the immediate future.[21] Women[22] and to a less significant degree teens and young adults[23] were more likely to send "grooming" SMS messages. Along the same lines, women were more likely to send emotionally based "grooming" messages.[24] This finding echoes that of Herring, who notes that in the world of e-mail, women are generally more "aligned" and supportive in their communications (Herring 2001).

Sex and sexuality are a part of the communication and often figure in commentaries of texting and text messages.[25] According to material reported by NUA, about one in four persons in Europe has sent or received sexually based text messages at some point in time (NUA 2002). While there are examples of this form of interaction in the corpus, it was so small as to be included in "Diverse other

Theme	%	Examples
Coordination	33	The car is done so we can get it at 4 (M, 52) Can you pick up the children at daycare (W, 30)
Grooming[26]	17	Good that it went so well with your math exam. You are smart, Love grandma (W, 58) Good night sex bomb (W, 35)
Answers	14	Yes, no, ok (2.3% of all messages) I have taken care of that (W, 31)
Questions	11	Have you caught any fish? (M, 40) Are you awake (M, 31)
Information	6	I found the sponge it was in the cork in the bottle (W, 28)
Commands or requests	6	Call me (1.5% of all messages) Remember to buy bread (M, 51)
Personal news	5	Mom has gotten [a] mobile. (W, 58) We are enjoying ourselves in the sun and good weather (M, 58)
Diverse other categories	9	Where are you? (1.4% of all messages) Thanks for the birthday present. love s___ (W, 17)

TABLE 7.1 The most common theme in SMS messages, Norway, 2002[27]

categories" in Table 7.1. Clearly, we must assert the caveat that this type of communication was underreported in the data. People are generally not willing to read intimate and revealing messages to survey canvassers over the telephone. Nonetheless these types of messages are striking in their absence, given their prominence in commentaries regarding texting. All told, only about 2.8% of all messages were jokes, and of these about two-thirds were scatological. Beyond coding the messages, the individual words were examined for the use of sexually related language. Less than 0.5% of the words had any direct sexual meaning. Quite often those that appeared in the Norwegian material were written in English and included words such as "ass" (once), "bitch" (once), "fuck/fucker" (one each), and "pussy" (once). Indeed there is only one example of the Norwegian spelling of "sex" but four variations of it in English: "sex" (once), "sexbomb" (once), and "sexy" (twice). We might conclude from this that the main threat to Norwegian posed by English is in the category of dirty words.

Obviously, sex and sexuality have their place in texting. It seems clear that respondents have withheld many of the saucier comments. Just as obviously, cheeky comments can be built up of innuendo and double meaning using otherwise commonly acceptable words such as "weenie." While it may be that informants were reluctant to provide their most salacious messages and while it may be that prudish Norwegians are behind other warm-blooded nations here, the assertion that sex and sexuality represent the dominant theme in texting seems to be overblown. The use of sexually related jokes and "sexual" vocabulary here seems meager, to say the least. Even if we assume an underreporting of an order of magnitude, there are still an extremely small number of both legitimate and risqué jokes. Based on this, we can question the popular image of texting as an open forum for freewheeling humor.

Mechanics of SMS Writing

We would expect to find a telegraphic writing style in text messages. The difficulty of writing and editing and the fact that writing is squeezed into the open moments of life would point in this direction.

The analysis seems to indicate that people use a vocabulary of coordination. The list of the most used words is full of pronouns (you, I, we) and prepositions (on, to, in, with) for both men and women.[28] Indeed, the particularly versatile Norwegian prepositions *på* (on/in/at/to) and *i* (in/at) make up more than 5% of all words used in the SMS messages examined here.[29] Verbs make a weak appearance in the 10 most used words, with commonly used variations such as "are," "have," "come," and, interestingly, "call." We find that *du* (you) is the word most used by men and the second most used by women. Hård af Segerstaad reports the same for a sample of Swedish SMS messages (2003a). Adjectives and adverbs, that is, words that modify or qualify nouns and verbs, did not make their appearance until far down the list. When they did appear they were words like "not," "now," "ok," and "soon." All of these words seem to point to people making agreements as to where they are going to meet, not, as some have suggested, bawdy innuendo.[30]

Going beyond which words are used, we can examine how many different words are included in a communication.[31] According to Yates, we expect the narrowest vocabulary, that is, use of the fewest different words, when speaking (1996). Writing, particularly formal writing — where a person proofs, revises, strives to eliminate repetition, and uses various reference works — theoretically allows us to draw on the broadest vocabulary. Using this measure, we find that e-mail seems to sit somewhere between formal writing and spoken communication (Yates 1996;

Baron 2003). In this context, SMS is quite different. Text production is actually quite slow, since the mechanics are far more complex than with alternative writing forms (Silfverberg *et al.* 2002). Thus, we might expect that the sluggish production process means a long time to think about alternative words. At the same time, a challenging form of text entry means we have to concentrate on the writing instrument and not the writing, making editing clumsy.

The analysis here shows that younger users employed a relatively limited number of different words in the production of their SMS messages (Ling 2003). Indeed, their language seemed similar to spoken Norwegian when considering the number of different words used. This seems to indicate that their text messages are written impetuously at the speed of thought, that the teens have a more limited linguistic reach than other groups, or perhaps both. In contrast to this, the mature adults seemed to draw on a broader range of different words in their messages.

Other dimensions for examining the material are the mean length (Figure 7.3) and the complexity (Figure 7.4) of the messages. The analysis shows that there is a significant gender difference in both the number of words per SMS message and their complexity.

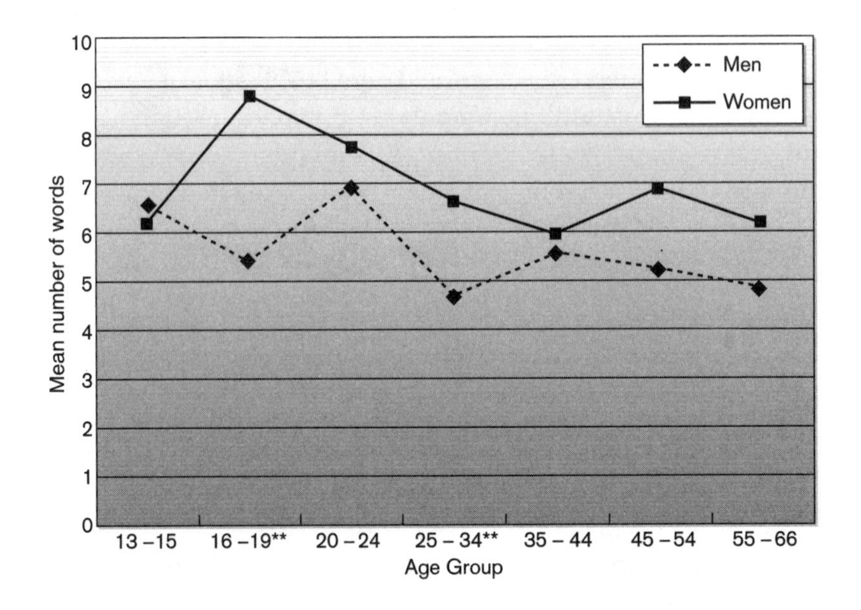

FIGURE 7.3 Mean number of words per SMS message by age and gender (** = Sig. < 0.01)

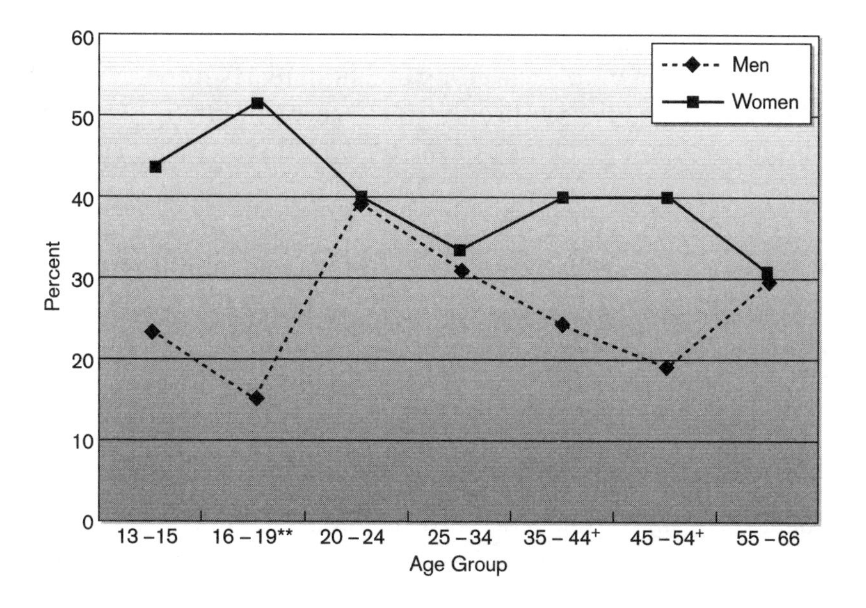

FIGURE 7.4 Percent of complex SMS messages, by age and gender (** = Sig. < 0.01, + = Sig. < 0.10)

Women generally write longer SMS messages. Indeed, the mean number of words per message for men was 5.54 words, whereas for women it was 6.95.[32] Women also seem to write more complex text messages. Men are more likely to write simple, one-clause/sentence messages, such as this SMS sent by a 15-year-old teen: *møttes klokken 5* ([We will] meet at 5 o'clock). The message is short, direct, and shorn of all unnecessary grammatical and punctuational niceties. About 66% of all the messages in the sample were simple one- "sentence" statements such as this. By contrast, about one-third of the messages were more complex in their construction, such as the following, sent by a 20-year-old woman: "*hi, yeah things are good with me you know! tried to call you just now but the number was not in use? do not know if I passed [my exam] or not*" There is a lot happening here. We see a report on her general situation, confirmation of her attempt to call, and finally a request for information.

It was more common for women to produce these complex messages. More than 74% of the messages sent by men were simple, one-sentence or one-clause messages, whereas only 60% of the messages sent by women could be seen as simple messages.[33] The data shows that girls 16–19 years old are particularly adept at writing complex SMS messages (51.61% of all their messages are complex). In an interesting contrast, boys in this age group are particularly oriented toward simple messages (84.85% are simple, while only 15.15% are complex).[34] Thus, we are presented with the fascinating contrast of relatively effusive teen girls and reserved teen boys.

Much has been made of teens' use of abbreviations in e-mail and SMS. Indeed directories have been published — and sold — listing a variety of acronyms and pruned spellings and emoticons. I examined the data both for abbreviations and emoticons. The analysis shows that in spite of all the discussion of this issue, only about 6% of the messages contained these forms of language.[35] The work of Doering in Germany suggests the same general finding (2002), though that of Ellwood-Clayton in the Philippines suggests the opposite (2003).[36] The data from Norway show that teen and young adult SMS correspondents are the biggest users of abbreviations/emoticons and that there is a rapid decline of use with age (Figure 7.5).[37] Women SMS users also include abbreviations and emoticons significantly more

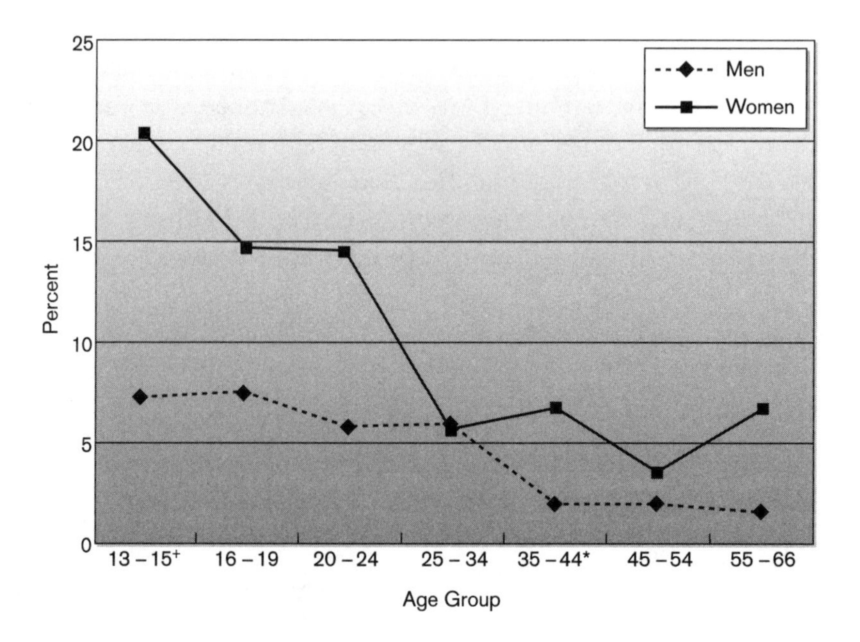

FIGURE 7.5 Percent of persons using abbreviations or emoticons, by age and gender (* = Sig. < 0.05, + = Sig. < 0.10)

than men.[38] Indeed, slightly more than 20% of women 13–15 years old used abbreviations in the SMS messages examined here. Only 3.5% of women 35–44 years old did the same. Women 35–44 years old and, to a less significant degree, those 13–15 years old[39] used more abbreviations than like-aged males. In group interviews of young adults, informants confirmed that the use of abbreviations was seen as an overly stylized form of language often associated with younger teens.

The analysis shows that the SMS messages from the younger users were more likely to have advanced capitalization and punctuation. Looking first at capitalization, we can distinguish three different levels of use: (1) no capitalization, (2) capitalization of only the first letter in the message (sometimes the default function in the mobile telephone), and (3) complex capitalization, including the use of capital letters in names, in proper nouns, and at the beginning of secondary sentences.[40] Approximately 82% of the messages had no capitalization, another 11% had only first-letter capitalization, and the remaining 7% had complex capitalization. There were no significant gender differences with first-letter capitalization. However, the SMS messages written by women were significantly more likely to have complex capitalization (4.9% for men vs. 8.5% for women).[41] Interestingly it is young adults aged 20–24 who are most likely to use capitalization in any form[42] and also most likely to use first-letter capitalization.[43]

Along the same lines, the analysis of punctuation compared those persons who used no punctuation in their message with those who used punctuation. The results show that young adults are also the most likely to use punctuation in their SMS messages.[44] Women use punctuation slightly more than men, but the relationship does not appear to be significant. Interestingly, punctuation seems to be used, instead of spaces, between words in some cases. Thus you might write, "*See you tomorrow,love Kari*" and use the comma to mark both the clause and the boundary between tomorrow and love.

Finally, we can examine the degree to which writers followed the form of traditional letter writing in their messages, i.e., including salutations and closings. Overall, there are relatively few messages that had either of these formulations. There were what we can consider simple openings and closings and more advanced or formal versions. The informal openings were often a chatty *hei* (Hi), followed in about half of the cases with a punctuation mark of some kind. In very few cases, the openings were more formal, including both a greeting and the name of the person being addressed.

The informal closings were either the name or initial of the writer or something like *Koz* (a stylized spelling of "hug") with perhaps an emoticon. The more formal closings used the common formulation of, for example, *Hilsen Jens* (greetings Jens) following a period.

In terms of the distribution of the salutations and closings, only about 10% of the messages had either an opening or a closing. The most common were the simple forms, with about 3.5% of the messages have a simple opening and 4.5% having a simple closing. The remaining 2% were distributed between messages with formal openings, formal closings, or both an opening and a closing. Thus, simple closings were most common. Amongst these, about half were the name or initial of the sender, and the other half were endearments, emoticons, or both. Again, it is women and teens that are in the forefront here. As in the case of e-mail, we find salutations and/or closings in the SMS messages written by women more often than in those written by men (Herring 2001).[45] Those under age 19 were also more likely to include these formulations in their messages.[46]

Written vs. Spoken Language

Returning to the more general question of the nature of text messaging. The analysis here seems to indicate that texting is a combination of verbal and written language. In another context I have referred to texting as a translinguistic drag queen, since it has features of both spoken and written culture but with enough flare of its own to catch your attention (Ling 2003).

Several elements cause us to think that texting is more like speaking than writing. First, we often find an immediacy to the communications. This can be seen in this impulsive message sent by a 15-year-old girl: *Eg kjeder meg* (I am bored). As with much spoken language, her statement is produced in first person present tense and a limited range of words. As with most spoken language, this message makes the assumption of informality and a lack of ceremony. Ironically, this varies by age, in that teen women use salutations and closings more frequently than others, giving their messages a tone of being a formally constructed letter.

Like speech, text messages are more ephemeral than letters. One can save text messages—within certain limits. Indeed, teens talk of saving the SMS messages that mark the establishment of romantic relationships. However, unless one transcribes them or sends them to others, it is more difficult to preserve them than, for example, words recorded on paper. To be sure, it is doubtful that in 50 or 100 years one will find a package of grandma's intimate SMS messages in the attic.

Further, statements such as that produced by the bored 15-year-old just cited are addressed to specific individuals. Unlike writing, which can be addressed to any reader who chances by, the vast majority of text messages are written with the intention of being sent to a single individual. Along the same lines, there is a high degree of personal disclosure in the text messages. That is, the sender and receiver have a high degree of insight into each other's lives and their immediate condition.

By way of contrast, text messaging is like writing, in that it does not assume that the interlocutors are physically proximate. Text messages are missives sent off to a person who is remotely located. At least in Norway, texting is generally more reserved than spoken language, in that it does not use adjectives or adverbs in any broad way. Indeed the analysis shows that there are almost no adjectives or adverbs among the 10 most frequently used words. And when they do appear, they are not the flowery type but rather are being used to arrange practical matters.

Like writing, text messages are editable to some degree, though the features for editing are more cumbersome than those found in PC-based writing. In addition, some mobile telephones provide the capabilities to capitalize certain words automatically, suggest spellings, and search through saved texts. Thus, several features of texting indicate that it is like writing.

Finally, there are features of texting that are ambiguous. For example, like letter-based correspondence, texting is an asynchronous form of communication. You send a message with the assumption that the addressee will eventually read it and respond when he or she gets around to it. It is assumed that you cannot necessarily command the attention of your counterpart in the same way you do in spoken interaction. Texting, as with e-mail and traditional letter writing, is not like a conversation, where pauses when taking turns are interpreted as impoliteness. This said, among teens the dialogues can take on the characteristics of a conversation, with the development of topics, the inclusion of opening and farewell sequences, and indeed the interpretation of pauses in turn taking. Teens in a group interview described how they actively use the pauses in sequences of text messages. With well-known friends they can reply immediately, but when courting a new boy- or girlfriend, responding too quickly or waiting too long to respond to the other's messages is a trickier affair. If they respond too quickly, they are perhaps seen as being too anxious, as with a person who sits beside the phone, waiting for it to ring, and answers on the first ring. In contrast, waiting too long can be interpreted as being to nonchalant. Thus, the timing of responding in this situation is a carefully thought-through part of the interaction.

Another ambiguity is the spontaneous nature of the medium. As with other forms of writing, we can edit a text message before sending it. Obviously, this allows us to monitor the content of the message and to avoid blurting out something better left unsaid. However, the ubiquitous nature of mobile telephony means that we can send ill-advised messages on the spur of the moment. Indeed, people describe "drunken" messages sent late on particularly "damp" evenings.[47] In this case, the sender is perhaps too uncontrolled in his or her comments, and there is no natural check on the ability to send the message across time and space. These have the unfortunate combination of being text based, archivable — at least in the short term — and spontaneous. It is in this case that one hopes that another characteristic of text messages kicks in, i.e., the assumption of privacy in the transmission. A text message is somewhat more private than a letter, in that it is password protected and because one can easily and permanently erase the message. However, it is also easy to copy and resend the message should one choose to do so (NUA 2002).

Gendering of Text Messages

What does all this tell us about the sociolinguistic nature of texting? At the broader social level, the results here indicate that, as in other spheres of language use, the culture of texting lives among younger women users. In spite of the fact that men were early adopters of mobile telephones (Ling 2000b), it is women, in particular younger women, who seem to have a broader register when sending text messages. They use them for immediate practical coordination issues and for the more emotional side of mobile communication. In addition, their messages are longer, have a more complex structure, and retain more of the traditional conventions associated with other written forms than those of men.

To set this into a broader context, women generally seem to have better interaction skills. In spoken conversation it is reported that women are better at the strategic introduction of topics of conversation (Fishman 1978; Treichler and Kramarae 1983), women are more accomplished at the use of rhetorical devices in order to maintain a conversation, and they are more likely to use various forms of critique and interpretation (Treichler and Kramarae 1983). And women use devices for adjusting and shifting the topic of the conversation, they manipulate the pauses, and they using various "grounding" devices, such as "mm" and "yeah," in order to facilitate the conversation (Sattle 1976). Historically this competence has been played out via letters (Krogh 1990), and it has also been extended to telephonic communication, in that women (and women's talk) is the medium through which parties are organized, familial news is sent, and remote care is communicated (Ling 1998b; Rosenthal 1985; Moyal 1992; Rakow 1988, 1992; Rakow and Navarro 1993; Tannen 1991). The material here seems to suggest that women are also more adroit "texters."

This is not to say that the writing of teen women is the polished prose of Margaret Mead, Toni Morrison, or Virginia Woolf. These are short and slapdash messages intended for immediate response. There is often a type of breathless, I-can't-wait-for-your-response nature to the messages. Nonetheless, in many ways the young women have the most respectful and sophisticated prose in their text messages. At least in this format and in this medium, the teens and teen women have control.

The Future of Texting

Texting is related very much to teens and young adults. There is, for example, no other communication medium in that is so intensely used by younger persons and at the same time so fundamentally ignored by their elders. At this point we can suggest that texting is a typically youth-oriented technology whose popularity is based on its relative low cost and youth-based mystique. Texting — like, for example, Gameboy and Nintendo devices — will remain a type of teen and young adult technology that resists adoption by other, "more serious," age groups. The second possibility is that today's teens and young adults will retain this technology — perhaps enhanced with photographs and sent across platforms as e-mail — and bring it with them as they move through life. That is, we can suggest a life-phase, or alternatively a cohort, model for the future use of texting.

It is tempting to suggest that texting will retain elements of both. Today's teens will carry the technology with them as a part of their repertoire of communication possibilities. Where older generations, when confronted with the need to communicate, perhaps go in search of a telephone booth, today's teens and young adults may just as naturally immediately consider using text messages. This said, youth and young adulthood are unique life phases (Frønes and Brusdal 2000). Mobile telephony and texting are particularly suited to the nomadic nature of these life phases. They allow spontaneity and a dynamic style of planning that is not as necessary in the more staid and routinized mature adult phases of life. Thus, while today's teens will carry texting with them into their maturity and while older age groups will increasingly adopt texting, its locus will continue to be among younger users.

The next question regards what will happen in the United States. While the United States generally leads, or is at least competitive with, other types of mediated communication, it seems to lag badly with the use of texting and indeed with the use of mobile telephony. There seem to be several issues here. These include interoperability, payment issues, and the popularity of alternative systems, most particularly instant messaging.

Unlike major portions of the world, the mobile telephone system in the United States does not use the GSM standard. Rather, there are several alternative transmission standards. The upshot of this is that it has taken time to sort out the ability to send text messages across systems. This, in turn, has meant that the general utility of the service is lower for the users than if interoperability were a fact. In addition, the tariff system in the United States does not embrace the idea of "calling party pays" (Robbins and Turner 2002). Rather, both the person placing the call and the person receiving the call share the cost.[48] From the perspective of the person being called, this is like getting an unwanted guest and having to pay for him or her. Thus, we lose the notion of a text message as a type of gift. Finally, many teens have a readily available alternative, most particularly instant messaging. It is common for college students, for example, to have free access to computers on campus and in dorm networks that include instant messaging. Just as the broad acceptance of the Minitel blocked the spread of the Internet in France, instant messaging blocks the need for the spread of text messaging for these persons. These issues have limited the spread of texting in the United States. The situation, however, is dynamic and will likely change.

Finally, we can ask whether there will be texting in 20 years. Again, the answer probably is yes and no. The wild success of texting has indicated that there is a role for asynchronous, mobile, and text-based communication. However, we suspect that as terminal design evolves and as more advanced and faster mobile networks come into use, today's text message system will be replaced by messaging systems that include broader functionality. Indeed this is already the case with the development of mobile telephones containing cameras and the ability to send multimedia messages (MMS).

It is likely that in 20 years today's teens will look back on texting just as my generation looks back on bell-bottom pants, wide collars, and paisley ties. When they send these reminiscences to their friends, however, they may well do so while sitting on a bus using glasses-mounted displays and virtual keyboards. And beyond reflecting on their own past, they may include the latest video of their child's retro rock band playing at some amateur festival.

8

CHAPTER

Conclusion: The Significance of Osborne's Prognosis

Introduction

In 1954 Harold S. Osborne, the recently retired chief engineer for AT&T, made the following prediction (quoted in Conly 1954, p. 88):

> *Lets say that in the ultimate, whenever a baby is born anywhere in the world he [sic] is given at birth a number that will be his telephone number for life. As soon as he can talk, he is given a watchlike device with 10 little buttons on one side and a screen on the other [see Figure 8.1]. Thus equipped, at any time when he wishes to talk with anyone in the world, he will pull out the device and punch on the keys the number of his friend. Then, turning the device over, he will hear the voice of his friend and see his face on the screen, in color and in three dimensions. If he does not see him and hear him, he will know that the friend is dead.*

The statement is both jarring and interesting. It is grating in its caviler attitude toward personal privacy and in its Orwellian prediction that we are perpetually available from cradle to grave. In addition, Osborne seems to be off target concerning the diffusion of mobile telephony.

It is interesting in that we have come a long way in realizing the prognosis. To set Osborne's comment into context, seven years earlier, in 1947, the world had seen the development of the transistor by John Bardeen, Walter H. Brattain, and William B. Shockley at Bell Labs in New Jersey. These devices had radically reduced the size, power requirements, and durability of all types of electrical devices. At the

FIGURE 8.1 An illustration of Osborne's "watch like" mobile communication device.

time Osborne made his statement, engineers were still trying to grasp the implications of transistors and imagining how they could be used.

Nearly 60 years later, portions of Osborne's prognosis have been realized. A modern mobile telephone need not be much larger than the device he described. In addition, modern mobile telephones allow for a host of additional functions, including the capture and transfer of pictures, graphics, music files, and other multimedia forms of information. There is the promise that the third generation of mobile telephony will include a type of video telephony — in color, though we must wonder if mobile video telephony will be any more successful than its landline predecessors (Ling 1998c).

From the perspective of a social scientist, I am also prompted to say that Osborne missed the boat in relation to the social consequences of the device. His statement is full of that wonderful 1950s awe of technology. There is the unstated sense that that we will seek out interaction with our friends and that this in itself is a positive thing. However, Osborne's prognosis betrays no real understanding of how the technology will be adopted in everyday life. He was not reported to have discussed the intrusive nature of his vision, nor did he speculate on the impact of the device on civil life. What, for example, will the mobile telephone mean for the way I live my life? How will the device change the way my daughters grow up? How will it affect the way they interact with their peers and the ways they organize their daily lives?

Obviously, it is not fair to make that demand of Osborne, since his work was to develop the technical devices, not to examine their social consequences. Nonetheless, his statement underscores what sociologist William F. Ogburn was, at about the same time, calling *cultural lag* — the idea that while various technical changes take place, their embedding in and domestication by the social structure lags behind (Ogburn 1950, p. 205):

> *When the material conditions change, changes are occasioned in the adaptive culture. But these changes in the adaptive culture do not synchronize exactly with the change in material culture. There is a lag which may last for varying lengths of time, sometimes indeed, for many years.*

This is also implied by Silverstone and Haddon's domestication (Silverstone 1995; Haddon 2001). That is, since the adoption of technology is a process that takes time, our social embedding of technologies lags necessarily behind their introduction. In addition, the way in which they are adopted and the uses to which we put them are not necessarily the same as what prompted their development.

Interaction Between Innovation and Social Institutions

The development of the transistor — and the subsequent integrated circuit — has been seen by many as a defining moment of the twentieth century. Indeed the revolution is not complete, since we are seeing the application of this technology to ever-more exotic items, ranging from superadvanced computing devices to birthday cards that do the job of singing "Happy Birthday."

What are the social implications of these developments? And, more germane for this book, what will be the social consequences of the mobile telephone? The previous chapters of the book have examined this issue at one level of abstraction. I have looked at the ways in which mobile telephony addresses everyday issues and is seen in everyday situations. Here, I want to look into some of the broader social issues of mobile communication. I want to place this into the classic sociological project of examining the interaction between technology and the social order.

History of Technical Innovation and Social Adoption

There are many examples of the interaction between technical innovation and social adoption. The most obvious of these is the industrial revolution, between approximately 1760 and 1840. New materials (iron and steel) and new forms of power (coal and the steam engine) led to the reorganization of work into the factory system.

These technical innovations arose out of various social contexts, and they resulted in a broad range of social changes. It was during this period that social observers witnessed the tremendous changes wrought by industrialization. There was the rise of wage labor. The extended family withered and reformed itself into the more transportable nuclear family. The church, once an institution wielding massive power, became a shadow of itself. Forms of governance and types of social movements saw fundamental changes. The division of labor took on finer and finer gradations. Labor organizations formed and were met in different ways by those who owned and operated the factories. Likewise, the legal, political, and social enfranchisement of various groups, most notably women, became a topic of pitched debate (as well as pitched stones). The rise of today's educational system was a child of the industrial revolution, as was the modern city itself. Thus, the interaction between industrialization and society resulted in a fundamentally reformed social landscape. The industrial revolution is, in many ways, the central epoch in the establishment of modern institutions. All other social changes must measure themselves against the changes of that period.

A second, more recent example, namely the history of the automobile, illustrates how one technology has become a central part of life for many persons. At the time of its early popularization, perhaps dating from the production of the model T Ford in 1908, the car owner lived in a world quite different from what we experience today. As with the development of the steam-driven industrial revolution, at the time of its development the automobile was superimposed on a preexisting social structure. Transportation was organized around either the railway system or the horse and carriage. From the perspective of the car owner there was a very sparse infrastructure. Less than 10% of the roads in the United States were surfaced during the first decade of the 20th century, gas stations were still in the future, and mechanics and garages were not readily accessible (Flink 2001). Indeed, a good part of the popularity of the model T owes to its durability and relative ease to repair and maintain. To top this all off, automobiles and automobile owners were often the butt of jokes and the victims of repressive legislation.

Social structure during the first decades after the mass adoption of the automobile was little affected by the automobile. The systems of retailing, work, and local interaction retained their "preautomobile" character. One bought clothes, food, and the other essentials of life in the same stores (Flink 2001, p. 3). If one lived in a city, the grocery store was around the corner and the drug store was not much farther away. According to the Lynds in their study of Middletown, almost half of the workforce lived within two miles of their job (Lynd and Lynd 1929, cited in Flink 2001). Thus, people could deal with many of their needs and responsibilities by walking.

However, in the period between World War I and World War II the situation began to change. Businesses began to move from the core of the cities to peripheral areas. Suburban shopping centers, such as the Country Club Plaza in Kansas City, Missouri, were built. The form of social interaction also changed. Rather than neighborhood-based visits, there was the "Sunday motor trip." There was the development of automobile-based restaurants, including the A&W drive-ins with their roller-skating "carhops" *qua* waitresses. There was also the development of automobile-focused motels,[1] drive-in movie theatres, and automobile-based tourism (Flink 2001).

The rise of the automobile also led to the development of new forms of social interaction. Where in the preautomobile period social interaction often focused on the neighborhood, after the adoption of the car what has been called "communities of interest" (as opposed to communities of propinquity) developed. Speaking of the interwar period, Flink (2001, p. 168) cites one commentator:

> *A new definition of a friendly neighborhood is apparent; it is one in which the neighbors tend to their own business. One householder explained: "We have nothing whatsoever to do with our neighbors. I don't even know their names or know them to speak to. My best friends live in the city but by no means in this neighborhood. We belong to no clubs and we do not attend any local church. We go auto riding, visiting, and uptown to theaters."*

Like the mobile telephone, the automobile extended the range of movement for youth and provided them with new, alternative forms of social interaction.

From our vantage at the dawn of the twenty-first century, it is clear that the automobile has changed society. In many cases, our work, shopping, worship, and social networks are inaccessible without the car (Hjorthol 2000). Local stores have been replaced by shopping malls. Traditional downtown areas have had to recalibrate themselves in order to appeal to the motorized public. The car has changed our sense of free time as it has increased our range of movement and opened up broader horizons that we can easily explore. In addition to the relatively immediate differences, the automobile has had its consequences for our personal and public economies. Building roads, erecting parking lots, and readjusting public space for the use of the car are common issues. In addition, the car uses enormous quantities of fuel and has resulted in local pollution and made its contribution to the greenhouse effect. Finally, according to Putnam, the car and the resulting suburbanization have contributed to the loosening of the social fabric (2000).

The widespread development of the automobile and the resulting suburbs has also given us the need for better real-time coordination (Townsend 2000). Automobile transport means that our extended families, our social circles, our work, and our forms of entertainment are flung across a wide geographical area. As noted in Chapter 4, the simple need to coordinate these elements is the niche into which the mobile telephone has established itself. Indeed, after the novelty of the device has worn off and it is seen as a matter-of-fact part of everyday life, the coordination of social interaction will likely be the most central consequence of the mobile telephone.

Sociology and the Role of Technical Innovation

In these examples, the development of technology arose out of a specific social context. In addition, there were a variety of social reactions to the innovation that are often moderated by Ogburn's cultural lag. Old institutions are changed. Bases of power arise as others are washed away. The innovations themselves are changed and molded as they are placed into the social terrain and as our understanding of their capabilities and their threats becomes more apparent. Following Silverstone and Haddon, the technologies are domesticated, in a broad sense of the word. They are turned to the purposes of those who control them, purposes that were not imagined by the early developers.

The analysis of society and technology has been a central issue in sociology since its beginning. Compte, Marx, Durkheim, Weber, and the other early founders of the discipline were often simply trying to describe the social impact of the dramatic technical changes they were witnessing. An integral part of the sociological project has been to describe the interactions between developments — such as those witnessed with the industrial revolution and the development of other technologies — and the way in which social groups and social institutions are affected. It is with these issues in mind that Durkheim developed his ideas of organic and mechanical solidarity, Marx examined the bourgeoisie and the proletariat, Weber developed his work on the processes of rationalization, and Tönnies contributed *Gemeinschaft und Gesellschaft.* It is also in this context that discussions about technical determinism, social determinism, affordances, and domestication of technology have taken place.

Thus, the themes here are not new. In our time, the discussion has taken on new dimensions. We seem to be encountering a new discontinuity, in the form of transistors, integrated circuits, the Internet, and mobile communication. Electronics and microchips are invading everything, from our cars, to our PCs, to our coffee makers, and they even are being regularly implanted in our pets. The development of the Internet and the adoption of mobile telephony have caused us to consider the degree to which we are on the cusp of a new social order.

Given the technical innovation, what are the broader social consequences? The Internet in particular has given rise to some wild-eyed speculation as to the social consequences of these developments. Some have suggested that the Internet will solve a wide variety of issues, ranging from international relations, to illiteracy, to saving the environment. The sense in these prognoses is that the open nature of the Internet means that it could allow the replacement of local communities with mediated communities of interest. In its most exotic form, others have suggested a type of Internet utopia that includes new forms of consciousness, the elimination of war and illiteracy, the resolution of the energy crisis, the saving of the environment, the elimination of the need for factories, and, in the words of one particularly fervent thinker, "crystallizing participatory democracy and resulting in a rich symbiosis of god and man, without the compulsion of power or law but by the voluntary cooperation of citizens" (Masuda in Kumar 1995, p. 15). In short, there would be no end to the potential of the Internet.

Back here on earth, a more realistic analysis indicates that the effects are far more sober (Katz and Rice 2002). Social scientists are starting to see that the Internet and the mobile telephone have a hand in the reformulation of the social order. However, these reformulations are not on the scale suggested by the transitions of the industrial revolution. While some institutions are in transition — arguably because of the influence of the Internet — we are not yet witness to dramatic changes.

By some estimations, the Internet has existed in the public imagination less than a decade. However, computerization and the trend toward distributed computing have been on the scene since the 1950s in one form or another. Thus, there has been time for some of the large-scale effects to take hold. However, concrete widescale changes in the major social institutions are difficult to find. To quote Kenneth Boulding, "The computer has brought us continually compounded interest at banks, easier airplane reservations, and a large quantity of unread Ph.D. theses" (Boulding 1980, p. 147).

The school experience of my daughters is quite similar to what I experienced. A teacher still stands in front of the blackboard demanding they understand the multiplication tables and the addition of fractions, just like what my fifth-grade teacher, Mrs. Reasoner — perhaps misguidedly — expected from me in 1965. Labor unions have conceivably lost some of their influence, but they still exist and still press their demands. The production system — and particularly the administration of the production system — has been made more effective in many ways, but wage labor is still very common. The nuclear family is perhaps split and reformulated with greater frequency, but two-generation households (i.e., parents with their children) are still the norm. While suburbs have become a feature of the landscape because of the automobile, the city as an institution is still familiar. While there have been shifts and changes, they are not on the same scale as during the industrial revolution. While the institutions are intact, we can ask how they are dealing with the shift toward individualism.

This book has attempted to look into the immediate effects of the popularization of the mobile telephone. These effects center on its use as a type of lifeline and in the coordination of everyday life, its rapid adoption by teens, and its intrusion into the public sphere.

How will these elements have played themselves out 50 years hence? Realistically, the major impact will be an improvement in our ability to coordinate activities. As suggested in Chapter 4, the mobile telephone represents the completion of the automobile revolution. The car allowed flexible transportation. But there was similar improvement in the real-time ability to coordinate movements, either in transit or in the cityscapes produced by modern transportation systems. We had to rely on telephone booths, preexisting agreements, and various fallback strategies. There was really no way to coordinate enroute since potential interlocutors were incommunicado. The mobile telephone completes the circle; it allows for this type of coordination. This is the true triumph of the system.

Beyond its function as coordination device, it is possible to imagine all types of fanciful mobile gizmos and technological developments, many of which are already making their appearance. Sending and receiving photographs, sound bites, and drawings, accessing the Internet, etc. are already available, as are certain context-sensitive services. Voice recognition, imbedded chips, glasses-based displays, and more precise, context-sensitive services all seem to be possible. These developments, however, are elaborations on the basic point of mobile communication. That is, mobile communication provides us with ubiquitous contact, both in social terms and in terms of access to information. Rather than requiring us to be at a specific geographical location, mobile communication means that we can communicate and have access to information wherever we are.

The chapters in this book have pointed out some of the immediate issues and examined these issues through the eyes of informants. However, as with the steam engine and the automobile, their interaction with society often resulted in turbulence on a broader scale. Many social formations that have arisen can be traced to the introduction of technical systems. Thus, we can ask whether the mobile telephone is a part of the more general information revolution. If so, what are its impacts beyond those described in the previous chapters? What effects will the device have on the role of the individual in society? Will the forces of centrifugal individualism offset those of society's centripetal effects?

Social Capital vs. Individualism

Social Capital

One area where the effects of the mobile telephone — and indeed of all forms of electronic communication — is being played out is in the area of what has been called *social capital* and its opposite twin, *individualization*. This discussion is a continuation of the traditional sociological project in which the interaction between a technological innovation and the workings of society is examined. If the mobile telephone contributes to individualization, it follows that the device also plays into our experience of social capital.

Social capital describes the web of trust and reciprocity that, in effect, binds the individual to society. Social capital is, on the whole, positive, since our ability to trust other social actors facilitates the functioning of society. The concept of social capital has existed in the sociological literature since the early 1980s (Portes 1998).

The first systematic use of the term was by Bourdieu (1985), who described it as access to various currently held and other potentially accessible resources that are based on group membership. A second approach to the analysis of social capital was provided by Coleman (1988), who posited a more ego-centered definition. According to Coleman, social capital consists of the resources available to an actor by virtue of his or her participation in a social group and that can be mobilized to achieve the individual's interests. A third approach is supplied by Putnam (2000), who describes social capital as the ties of reciprocity, trust and expectation that we carry with us and that are built up through formal and informal social interactions. In some ways, this is the most satisfying sociological definition, in that it finds its orientation not in the individual but in society. Putnam (2000, p. 19) notes:

> *Whereas physical capital refers to physical objects and human capital refers to properties of individuals, social capital refers to connections among individuals — social networks and the norms of reciprocity and trustworthiness that arise from them. In that sense, social capital is closely related to what some have called "civic virtue." The difference is that "social capital" calls attention to the fact that civic virtue is most powerful when embedded in a dense network of reciprocal relations. A society of many virtuous but isolated individuals is not necessarily rich in social capital.*

As people interact with their local milieu, they develop a repertoire of interactions and a sense of identity based on the interactions. There is, in effect, a type of synergy, in that they can draw on the relationships in other contexts and, thus, the form for the interaction extends beyond the specific into the more general.

Putnam has examined material describing a broad range of civic and social participation and, in turn, the fate of social capital in the United States. His analysis includes participation in formal organizations, informal social interactions such as dinner parties and local neighbor coffee klatches, etc. Generally, he finds that there are both cohort and life-phase effects that, in sum, have led to a reduction of social capital in the United States.

The cohort effect arises from the fact that members of the pre–World War II generation are slowly passing from the scene. As they do, they are taking with them an extraordinary level of social solidarity and engagement. It was this generation that experienced the melding effects of both the Depression and World War II. This generation was central in founding and supporting many community-based organizations, including parent–teacher associations, bowling leagues, and an array of social and civic groups. Their commitment to voluntary communal activity is difficult for other generations to match. The children of the baby boom, generation X, and the other cohorts have not mustered the same passion for communal activity, and thus the passing of the prewar generation has resulted in a sum loss for society on this front.

Beyond the passing of a particularly engaged generation, Putnam points to the effects of suburbanization and the rise of TV culture. In his estimation, the geographical growth of cities and the insularity provided by TV mean that we are more isolated and less disposed to participate in social interaction. The fact that social interaction does not take place over the backyard fence but, rather, in a physically remote location acts as a barrier to the development of social capital. In the same way that the distances imposed by the growth of the suburbs hinders social capital, the time and attention afforded to TV are a barrier. To the degree that we use our free time to watch TV, we are not available for other types of social interaction.

Social capital is not, however, only a positive force. The dark side of social capital is that it can result in the exclusion of outsiders, excess claims on group members, restrictions on individual freedoms, and a downward leveling of norms (Portes 1998). The downside of social capital can be seen in cronyism and, in the case of teens, the formation of cliques that are more or less hermetically sealed.

To the degree that gangs and cliques press their uniqueness as being more important than the common good of the whole, there is a dysfunction. This is a difficult balance, particularly for teens. As we have seen, adolescence is a period in which the development of identity is an important issue. The peer group — be it a clique, a gang with perhaps quasi-formal induction requirements, or a slightly more amorphous grouping — is often a central experience for a teen. However, when membership in such a subgroup takes precedence over participation in other broader groupings, such as school or a local society, the teen is closed off and does not have the opportunity to build up the broader web of relationships that help to form society.

The Institutionalization of Individualization

The opposite of social capital — and its proclivity for sociability — is individualization. Instead of the coalescing of social groups, there is a centrifugal tendency and the push toward atomization.

According to some, the relevance of social institutions is being put into question (Beck and Beck-Gernsheim 2002; Rasmussen 2003). In a continuation of Weber's rationalization of society, we are seeing the increasing ability to focus on the individual. Information technologies are, in many ways, the motor of this process. Rather than dealing with classes of persons such as all teens or all childless married women over 45, information and communication technology (ICT) allows a yet more finely tuned focus. We are judged not by our ascribed or even our achieved status, but rather through which market segment we inhabit at the moment.

I need only look at the reading recommendations I receive from Amazon.com to see the ways in which I have become distinguished in my individuality. I have ordered a book on the American desert by Wallace Stegner, a small volume on the history of mobile telephony, *Constant Touch* by Jon Agar, and *The Automobile Age* by James Flink. This has allowed Amazon to place me into a specific category in which my soul mates are perhaps a student in California, a homemaker in Bristol, an academic in Berlin, and a journalist in New York. That is, I am a data point in a dispersed and unconnected aggregate of individuals who have the same profile according to Amazon. I will never meet these others. Even if I were to meet them, we would not likely hit it off socially, in spite of the fact that we have the same taste in literature. I doubt that we would become bosom buddies and establish a club or social movement together. However, in this theoretical grouping, we see the increasing ability to treat persons as individuals. At the same time, we see the atomization of society. That is, we are not the members of a class, a family, or another type of grouping so much as we are conceived of as a set of individual interests.

The other dimension to this development is that we place ego before collective. Beck and Beck-Gernsheim see this in the historical development of marriage as an institution. It has changed from being determined by material tasks in the preindustrial society to being a material and emotionally based institution in industrialized society to being more exclusive an emotional bond between two relatively independent individualized economic actors (Beck and Beck-Gernsheim 2002, pp. 72–73). Here is one institution that, according to some, has progressed from a form of common solidarity — albeit imposed — to a situation of individualism. According to Beck and Beck-Gernsheim, this is the general trend in society. Indeed, they describe this as an indication of the somewhat oxymoronic institutionalization of individualism.

The use of the word *individualism* is unfortunate, in that it carries with it the sense of ego and "me first." While this form of egocentrism fits, the issue being described here is broader. It is not an individualism that is being forced by the actors but, rather, an individualism that arises out of the direction of the social order. In essence, the tendency in contemporary society is to handle persons as individual cases rather than as members of any particular class or category. This can be seen in the tailoring of insurance policies, education, career development, and the marketing by online booksellers. From the perspective of the individual, there is a freedom associated with this development, but there is also the need for the individual to do the job of placing him- or herself into the social order. Where before, the individual had a relatively ascribed status as a worker, a socialite, a student, or a local resident, now all of this has, to a much larger degree, become an achieved status (Rasmussen 2003, p. 10):

> *The point is that the individual's life is lived through a huge number of small decisions about situations that before were almost culturally given. Now we negotiate, decide, and renegotiate to overcome the content and the complexity. The faster, the more fluid, and the more continually we can maintain our overview — preferably from hour to hour — the better off we are [author's translation].*

All of this means that where before many of these issues were settled once and for all, they are now being continually renegotiated as we move about in our daily life and as we move through the various life cycles.

Role of ICTs in the Fostering of Social Capital/Individualism

The Internet

Information and communication technologies (ICTs) are clearly elements in the development and maintenance of social capital and individualism. Via the Internet, I am able to send e-mails to colleagues and friends literally around the world, thus maintaining social capital. In addition, the open nature of the system means that people with whom I have no preexisting relationship (along with a seemingly endless number of spammers) can reach me and exchange ideas. In this way, the Internet supports the development of social capital.

While Putnam focuses on several causes for the disruption in social capital, he is less sure as to the general effect of the Internet, and he offers little or no discussion of mobile telephony (Putnam 2000, pp. 170–180). Internet surfing, like TV, can play into individualism and can isolate people and monopolize time that might otherwise be spent interacting with their immediate social sphere. On the other hand, it is through the Internet that communities of interest can develop and grow. The technology allows people to organize and orchestrate activities across broad distances.

Beyond Putnam's discussion of the issue, several other researchers have contributed to this discussion. Robert Kraut, Sara Keisler, and their collaborators studied some of the social psychological effects of Internet use among new users and outlined the possibility that use of the Internet can lead to social isolation and depression (Kraut *et al.* 1998).

A contrasting finding as been reported by Katz *et al.* (2001; see also Katz and Rice 2002), who found that the social aspects of computer-mediated interaction is an important element in holding together task-oriented groups. Beyond this, they found that those with the most Internet experience seemed to have higher levels of social interaction. A similar finding arises out of Kavanaugh and Patterson's analysis of the Blacksburg Electronic Village. In this case they found that the longer people were involved in the Web community, the more likely they were to use the system for the development of social capital (Kavanaugh and Patterson 2001, p. 505). While this finding points to the leavening effect of Internet-mediated social interaction, there is also the suggestion that such communities thrive where there is already extensive social commitment beforehand. This has been described as a type of "the rich get richer" argument. The point here is that the Internet does not generate social capital; it only facilitates greater levels of social capital where there is an existing structure.

As already noted, the Internet is not a single service but, rather, is composed of functions, such as the World Wide Web, where we can gather information, download files containing music or pictures, maintain a blog, or simply "surf" from one point of interest to another. In addition, there are several more "social" functions where one engages in either person-to-person interaction (e-mail, instant messaging) or various group interactions (chat groups, list servers, MUDs, MOOs, or Usenet). While surfing the net is often our image of Internet use, it is indeed e-mail that is the most used function on the Web (Mante-Meijer *et al.* 2001).

With these distinctions in mind, Franzen suggests that surfing results in a small but significant reduction in the degree to which we participate in social activities with friends. However, if people reported using e-mail extensively, they also reported a larger social circle (Franzen 2000). Similarly, Anderson and Tracey — who followed a group of individuals in a panel design[2] — found that the use of e-mail increased from time 1 to time 2. Thus, the more experienced people were on the Net, the more likely they were to use e-mail, the most firmly established social channel on the Internet in Europe (Anderson and Tracey 2001). Whereas Franzen as well as Anderson and Tracey suggest that e-mail varies with greater sociability, Nie questions the link between these two concepts (Nie 2001). Nie asserts that the Internet does not cause greater sociability; rather, those who are already sociable, i.e., those with more advanced education, who are economically secure and who are in the more active phases of their lives, are the most likely also to adopt the Internet. According to Nie (2001, p. 429):

> *Internet users do not become more sociable because they have used the Internet, but they display a higher degree of social connectivity and participation because they are better educated, better off, and less likely to be among the elderly. The point is that they report a steady decrease in social interaction with family and friends the longer they are using the Internet, in terms of either weekly hours or years online (despite the fact that they also report an increase in e-mail use!).*

Finally, Katz and Rice, using data from a national survey in the United States, report that Internet users did not differ from nonusers in terms of political activity or in other forms of sociability, such as letter writing and calling. However, in sum, Katz and Rice differ with the estimations of Nie and those of Kraut *et al.* in that they see the Internet as a positive force in the development of social capital. They note (Katz and Rice 2002, p. 337):

> *Rather than a technology of isolation and loneliness, the Internet is a technology through which social capital can be created. Its capability may be entirely potential and not used. But in many cases, it draws people into contact with others to create shared resources and communal concerns.*

Thus, the discussion as to the social consequence of the Internet has gained nuance as it has progressed. Starting with rather dreamy estimations of the Internet's potential, the discussion has increasingly brought the complexity of sociability into focus.

The Mobile Telephone

The rise of the mobile phone also has implications for the discussion of social capital/individualism. As we have seen in the previous chapters, the mobile telephone is a technology that can help to sustain social networks. Social groups can be more tightly integrated via the use of the device. In many respects, the mobile telephone is like the traditional landline telephone, in that it allows us to maintain contact with our social network (Putnam 2000, pp. 166–169; Wellman and Tindall 1993; Wellman 1996; Kellner 1977).[3] However, the mobile telephone has an advantage over the traditional telephone, in that it is more personalized and allows quick and direct access to others regardless of location.[4] Further, when calling on a mobile telephone, we call an individual, not an address or a family as with the traditional telephone. That is, I call directly to my wife or — on that day my daughters finally get a mobile telephone — I will be able call them directly and individually to find out when they had planned to come home to dinner. The device allows me a very direct and indeed a very individualized medium through which I can maintain social networks.

Our analysis of Europeans' use of the mobile telephone shows that there is a significant covariance between its use and informal social interaction. That is, the device is used — particularly by teens and young adults — to arrange an array of informal social activities, such as café visits and meeting with friends. (Ling *et al.* 2003). This finding is based on an analysis of the *e-living* database that covers six European countries. Interestingly, material in that analysis indicates that the mobile telephone is not an integral part of participation in formal social interaction in clubs, interest organizations, etc., nor was it a significant factor in the organization of the very closest friendships. Thus, the mobile telephone seems to be a type of middle-range technology.

It is a technology that supports the intense, informal, and local social interaction that is characteristic of the adolescent phase of life (Smoreda and Thomas 2001a).[5] The mobile telephone is also a totem for teens. It provides social integration at the symbolic level, and it provides the individual with a sense of self (Pedersen 2002; Cohen and Wakeford 2003).

In these ways, the mobile telephone nurtures at least some forms of social capital. This is not to say, however, that it supports all forms of social capital development. The same analysis cited earlier did not show a significant interaction between the mobile telephone and participation in formal social groups.

Thus, the mobile telephone allows us to be in touch perpetually with others. We can send messages or call from wherever there is coverage. The interaction can be initiated quite spontaneously.

One line of thought here is that this intensity of interaction — the mere weight of the interaction — serves to weld the social group together, be it a family or a group of teens. We know, on a nearly perpetual basis, where our friends or partners are, we know their immediate plans, and we have a general sense of their current situation. The couple knows each other's whereabouts and has the ability to rework plans as needed.[6]

Teen mobile phone users might know, for example, that their friend has just left home en route to a second friend's home, and that there is a general plan to continue on to a party at a third location. They may know that their friend will be wearing a certain type of clothes and incidentally that she is frustrated with her parents since she has to be home by 11 in the evening. After a short interlude, the individual may receive an update indicating that the party has been judged to be a potential dud. Thus, there is a renegotiation and the decision to meet at a café instead to consider which movie to see. We see here a type of continual interaction. Agreements are made and frequently readjusted. Background information is exchanged. The situation as well as the well-being of the members of the circle are monitored frequently. Rivere and Licoppe refer to this as a "connected" form of interaction (Rivere and Licoppe 2003).

With this style of interaction, a person's very "presence," in a virtual sense, results in a type of awareness and empathy that is developed and maintained through short bursts of communication.

The alternative is emotionally based interaction in which coordination is not the point. Instead, interaction at a more expressive level provides the bond. Rather than a series of quick exchanges, social bonding arises out of more extended interaction. In this variation, we set aside long periods to talk. During the conversation we report on our small and large victories and problems. In addition, we allow the others a chance to empty themselves of their issues and to share their insights.

Clearly, a relationship is not an either/or situation in which the interaction is all in either 2-second MTV-like clips or long Bergmanseque dialogues. Rather, there is a mixture. In one cycle, there is the need for intense contact; in the next cycle, there are short messages and coordination. The maintenance of a long-term interaction with another probably involves both parts of the cycle.

Nonetheless, reliance on a relatively quick and cheap form of interaction means that we might dispense with the longer and more ponderous process of a personal visit. While this may be acceptable in the context of a same-age peer group, it can also result in superficiality. If teens can dispense of their "Mother's Day responsibilities" with an SMS message instead of a personal visit or even a call, is this a tightening or a loosening of the social bonds? While the form of celebrating the day has been observed, we can question the formulaic way in which it has been carried out.

The other side of the coin is the capability of mobile communication to contribute to individualization in society. Indeed, the mobile telephone represents the increasing individualization of mediated communication. The device — the actual physical handset — is owned and personalized by the individual. Our friends are listed there along with the record of whom we have called. Teens save SMS messages from their paramours. We have our own distinctive ring, our own distinctive telephone cover, and our own distinctive logo. Increasingly we have our own photographs, songs, and even small electronic change purses onboard our mobile telephones. Mobile telephony is used to coordinate our individual actions and movements vis-à-vis other individuals in the social group. The point is that the device is our personal communication link. This is new. Before this, the landline telephone was a device that was shared by all those in the household, if not the neighborhood.

In this context, I recall a group interview in a small village in the Hardanger fjord area of western Norway. The topic was how harried one of the informants felt with all the telephone calls she received via the landline telephone. The woman said, in quasi-exasperation, that sometimes she simply wanted to unplug the phone to avoid the disturbance. At this point, an elderly respondent reacted sharply. It seems that his parents were among the first persons in the village to get a telephone. Given the communal nature of the village, the device was a link with the outside world. It was there in case of emergencies, it was used by neighbors to communicate with remote kin, and it was, in some respects, seen as a common resource. Given this history, the man could not understand that someone's individual frustration would outweigh the need for common access. Since the telephone was placed into the category of being a common resource, it was not an optional accessory. The fact that the device was potentially a life link and was in many ways the medium through which people maintained contact, it was unquestioned that it always be in service. Thus, the idea of an on/off button, as in the case of a mobile telephone, did not fit the paradigm of the older informant. It also underscores the way in which individual prerogative has begun to play into an area previously more socialized.

Cleary, the mobile telephone represents a sea change here. While one can point to similar socialized-use models, such as one finds in Bangladesh with the Grameen Phone experience, the common experience in many countries is that the mobile telephone moves toward fulfilling Osborne's comments. It goes a long way toward being the individualized communications medium that provides us with continual access to others. Not only are we individually available, but we are also becoming situationally available. Mobile communication systems allow us to determine the rough location of our friends, a nearby taxi, or a variety of other persons, items, or services in real time. Beyond being a personalized communication channel, the mobile telephone seems to be an important self-reflexive symbol (Pedersen 2002). That is, the device is not just our link with the world; it is a way in which we develop a sense of ourselves.

Ad Hoc Networks

An interesting twist in the social capital/individualization discussion is the potential for mobile telephones to support ad hoc networks. Obviously there are various applications in which the individual joins a group or organization where the point is to receive various types of dynamic updates (Rheingold 2002). Groups of teen girls in the United Kingdom who are intent on following the location of Prince William or other celebrities, as well as motorists who share their knowledge of radar speed traps along the highways, come to mind. In these cases, there is a common — albeit egotistical — interest in certain types of information. The other point is that you "join" this group; it is not a place where you simply drop in, as, for example, with chat groups on the Net. Rheingold has called these *smart mobs*; he suggests that they are a type of self-organizing, peer-to-peer social grouping that comes and goes according to need.

While these applications often focus on issues such as attempts to outfox the local police or to update the gang on the whereabouts of a rock star, this form of social organization can also turn to political ends. The actions in Manila with the revolt against Estrada, the Seattle and Gothenburg actions against the World Trade Organization, and a growing list of political actions show that the mobile telephone can be used in these situations.[7]

These social aggregates function as a unit so long as there is a shared ideology and a common sense of strategy, and so long as there is a focused and easily communicated form of interaction. These elements, along with the use of the traditional "telephone tree" — and its implied hierarchy of command — enable these forms of collective action.

From a sociological perspective, the revolt against Estrada arose out of a generally communicated national crisis, i.e., the sense that Estrada was corrupt. The general background tension meant that people were disposed to action. Given the generally tense situation, the decision by the Senate in the Philippines to ignore certain information in the case provided a clear focus for the actions. Beyond these ideological elements, there were the strategic matters to be dealt with. The specific location of the protest — Epifanio de los Santas Avenue — has symbolic value, in that it was the location of the 1986 "people power" mass protests against the Marcos regime. This location is known locally as "Edsa." It is a short, well-recognized name and is easy to send as an SMS message. Thus, there was a common ideology, a triggering event, and a common and easily communicated strategy.

The "marching orders" were effectively spread to the participants via SMS and then sent further to others using the list of contacts registered in their mobile phones. In a nicely turned phrase, this is what Ronfeldt and Arquilla as well as Rheingold calls *swarming* (Rheingold 2002). Given this clear focus, the mobile telephone served as an easy channel through which participants could be mobilized. Had any of these three elements been missing, the protest might well have gone flat.

As noted above, this and other, similar, actions have been successful to the degree that there was a common focus for the actions and no fractionalization in the movement at either the ideological or the strategic level. This critical point is often missed. According to Gamson, the anathema of social protests is fractionalization. His analysis shows that protests that are able to maintain a structure and that can avoid internal squabbling are the most successful (Gamson 1975). In the case of the Estrada protests and the WTO protests, these threats to the solidarity of the protesting groups have been held in check. Indeed this may be due to the short-lived nature of the protests.

The same dynamics that allow the movement to gain steam quickly can also provide factionalism a foothold. Insofar as broad masses can be rallied against a major foe, factional groups can also assert their agenda. We can speculate that the mobile telephone will allow for certain types of social action, but that these are necessarily limited by the ability of the movement to maintain a focus on the activities at hand and to avoid the establishment of alternative strategies or alternative ideological approaches to the problem.

Finally, these ad hoc groups play into the more general picture of mobile communication. On the one hand, their rise has been one of the startling issues associated with mobile communication. In addition, this form of interaction has potentially far-reaching effects. On the other hand, these forms of social interaction are relatively uncommon in the broader sweep of mobile communication. As I have tried to show, it arises out of particular social situations. While these social formations are interesting in that they are exotic and powerful, they are also transitory. They are not the everyday, routinized form of communication that forms the basis of social networks. Celebrity spotting or speed-trap evading are either interesting free-time hobbies or evasive strategies for avoiding official sanctions. The protest movements are mayflies that seemingly pop up and evaporate in the same news cycle. In the meantime, mundane telephonic communication continues. Parents call to coordinate delivery of children to soccer, teens girls send SMS messages comparing impressions of the new guy in class, carpenters order lumber, and businesspeople compare notes on the boss's attitude at the last sales meeting. While swarming gets the headlines, these functions constitute the mass of telephone use.

Another question is the degree to which these movements contribute to social capital. Obviously, those who participate are in contact with other protesters or celebrity chasers. There is the intrigue of the protest and the rush of participating in a broader social movement. However, we can ask whether this is a momentary experience or a broader social movement. Does it compare to the day-in, day-out intensity of other mass actions? However seriously felt by the individual, the protest or the celebrity hunt may be a type of momentary engagement that is juxtaposed against the other issues being faced, such as the pursuit of a career or the completion of an education.

Virtual Walled Communities

The mobile telephone seems to allow the development of some forms of social capital. It also plays into the institutionalization of individualization, to use the phrase of Beck and Beck-Gernsheim. How will these two seemingly contradictory tendencies play themselves out? Will the mobile telephone result in a flowering of the social sphere, or in the retreat to a balkanized social clique? Will the mobile telephone result in a society where the threshold for contact is lower, thus giving us access to a wider circle of friends, or will it intensify our circle of friends and provide us with a stronger internal solidarity? Will the mobile telephone turn us into members of a type of walled community where we interact only with a limited circle and routinely exclude others? Will the device lead to a more "postmodern" interaction in which we carry out — or are exposed to — a series of what might be seen as semicompleted banal interactions without broader context? Or does the mobile telephone actually allow us to connect spontaneously and instantaneously with others in our social circle when we would otherwise be simply staring out a bus window wishing we were someplace else? In sum, does the device add to or detract from the social capital of society?

On the one hand, the mobile telephone allows nearly unhindered access to our local social network. Indeed, it is the local peer group that is the locus of mobile telephone interaction (Smoreda and Thomas 2001a; see also Ito 2003). At the same time, using the mobile telephone sets up a barrier between ourselves and our physical situation. Engagement in the telephone call closes us off from other, copresent activities (de Gournay 2002).

The mobile telephone can strengthen our ability to maintain and elaborate contact within the immediate and intimate spheres. We can exchange messages and have an advanced and ongoing dialogue with like-minded people, regardless of when it is or where we are. The model airplane enthusiast can, via SMS, be engaged in a deep discussion of the pros and cons of various motors or wing forms — along with the appropriate insults and jibes — while sitting on a bus next to a star-crossed teen boy who is hopelessly in love for the first time and who is anxiously awaiting the next missive from his Juliet. The ability to be so actively engaged in remote affairs while in a public space is new. The older sense of community is being transformed and reconfigured (Green 2003). This is the ability to, as it were, have a foot in both the here and now as well as the there and now.

We can speculate that the intense interaction of the in-group can have a chilling effect on the ability to engage in more superficial and peripheral social relationships. Thus, the teen girl described earlier was so busy sending text messages to organize her meeting with her friends at the local café that she was unavailable for small talk with others at the bus stop. To the degree this is true, the mobile telephone supports the development of cliques. That is, the device allows for an intensity of interaction within the group that can easily tip over into what Gergen calls a type of "us against them" mentality (2003). Gergen suggests that this fragmentation can take place in subtle ways. Interaction with the group can monopolize an increasing amount of time when it is maintained remotely. We can attend to the group needs in the odd moments while we await a dental appointment, while we are waiting for the children's piano lessons to end, or at the bus station when we are — at least nominally — available for some superficial social engagement. In addition, the parochial concerns of the group can be magnified and thus be played out with more intensity than otherwise. At the same time, we become unavailable to those who are copresent. We do not engage in the small pleasantries and idle chat with our fellow waiting room colleagues. Thus, we can wonder how the ubiquitous access to our social group will play out society's total social capital.

In their application of Richard Sennett's ideas on the growing incivility in society, Rivere and Lecoppe bring out this same theme (Rivere and Licoppe 2003). They note that as communication forms become more focused on intimacy and personalization, we find a similar development of incivility in the public sphere. Personal and emotional interaction arises out of the increasing ability of individuals to interact with those of a similar mind, that is, communities of interest, as opposed to communities of propinquity.

In the midst of all this, what is holding society together? Where do we find the common ethic on which our willingness to interact is based? From Durkheim, we know that copresent ritual is a central element in the establishment of a common ethic. Writing as he was during the dawn of mediated communication, his focus was clearly on nonmediated social interaction. A central point in his work is that a social bond is formed when a group carries out a ritual, when in his words, they are "uttering the same cry, pronouncing the same word, or performing the same gesture in regard to some object that they become and feel themselves to be in unison" (Durkheim 1954, p. 230). He meant that the participation in the group, and the obeisance to a common focal point, is the mechanism through which we develop a sense of social solidarity, a sense of *nomos*. It is through common ritual that the community becomes conscious of itself and that the individual is aware of his or her participation in the whole. That is, there is a melding of the individual and the collective. Following on this, Collins has noted that there are several elements prerequisite for the Durkheimian ritual, including (1) face-to-face presence of the group, (2) a common focus of attention, (3) shared emotions, and (4) nonpractical actions carried out for symbolic ends (Collins 1994, p. 206).

As with Durkheim, Collins asserts the need for copresence when establishing the group ethic. Again, there is the need for the immersion of the self into the broader social flow. Simply watching the same TV program or participating in the same chat group is not enough. There is the need that we give ourselves over to the social, that we directly experience others doing the same, and that we know others have seen us joining in the group. This can take the form of prayer in a church, watching a football game, chanting slogans at a political rally, or being initiated into a fraternity. In each case, the individual is a part of the broader whole. The ethic can be prepared for and maintained via mediated channels (Ling 2000a). Nonetheless, we have to touch base occasionally, since without a shared and focused common meeting the common ethic dissipates.

How will the mobile telephone play into all of this? Will it allow the development of a shared ethic across a broad group of people, or will it enforce the bonds of the small group? Will the mobile telephone become a way of supporting a broader, more differentiated social circle, or will it allow us to develop the ritual solidarity of a small sphere of like-minded persons? The evidence, albeit quite preliminary, seems to point in the direction of the latter.

Miyata *et al.* report that in Japan so-called Web phones are most often used to send short and quick messages to friends who are nearby (2003). That is, they are being used to reinforce the strong ties of those who are physically and socially close. They are not being used to gather information about social issues, nor are they being used to cultivate the weaker ties. Something of the same is reported by Park: The mobile phone in Korea is being used to strengthen existing social connections while limiting people's ability to establish new social connections (Park 2003). Rivere and Licoppe describe what they call *connected interactions*, that is, a "seamless web of interactions, whether face-to-face, phone conversations, or multiple forms of messaging" (Rivere and Licoppe 2003; see also Licoppe and Smoreda 2003). They go on to note that this type of running interaction is limited to only the very closest persons in the intimate sphere. The development of electronic messaging has facilitated the development of these relationships in which a person has a running sense of the other's location and situation. Thus, there is a type of remote presence. Paradoxically, they point out that from the perspective of the individual, this is a civilizing effect in an "uncivil" world. That is, the specific and literally unceasing relationship to another intimate provides the individual with an oasis in an otherwise difficult world. From a social perspective, however, this represents a withdrawal from the public into the private. These researchers describe the balkanization of social interaction. There is the sense that "walled communities" are being formed because of the mobile telephone.

In some respects, the earlier chapters have taken up the same theme, particularly the analysis of security and microcoordination. Both of these functions take place in the context of the small group. We feel security in the ability to call a friend or to be in contact should something arise. By the same token, the ability to coordinate and indeed to microcoordinate means that the interaction is necessarily limited to a limited group of persons. In both cases, the interaction is generally within the sphere of the immediate group. The ability to harmonize the flux of everyday life and the ability to rely on sanctuary — as needed — in the intimate sphere are strengthened by our use of the mobile telephone. In a sense, the device allows us to draw on the latent solidarity of the group, a solidarity that is perhaps built up in other ritual contexts. However, our reanimating of the group solidarity takes us away from other opportunities to establish new ties in our colocated situation.

In "The Tragedy of the Commons," Garret Hardin discusses the ways in which common resources are progressively used up by the individualistic rationality of participating actors (1968). If we consider the public sphere as a type of commons, the mobile telephone brings up two issues. The first is what we might call a type of audio pollution of the public sphere because of the increasing number of mobile telephone calls. This may be a transitory problem. Just as noise pollution from other sources was seen as a problem during certain phases of technological development, the developments themselves and our ways of dealing with them have resulted in a code of behavior that soon removed the problem. SMS, for example, is one adjustment, as are other developing forms of courtesy. Here it seems that the sense of the commons is being reasserted through various adjustments.

The other issue is the withdrawal from the public sphere. As Jane Jacobs noted, the thing that makes the public sphere vibrant is the continual contact with unexpected forms of interactions (Jacobs 1961). Not all are pleasant, and not all are sought. Nonetheless, there is vitality and a roundness that arise from our interaction with a variety of others, no matter how perfunctory. Seeing the legless beggar, watching a street musician, giving directions to the tourist, and seeing the exotic hair color and shockingly mismatched clothes of the older woman are all elements that inform us as to the mood and spirit of our local situation. Better this than some Stepford-like existence in which all is neatly tucked into the same pattern and alternatives are not only frowned upon, but eradicated.

At a milder level, being part of the public sphere means that we are available to tell another passenger on the bus that this is the bus stop they asked about. It means being able to ask another what the time of day is or to comment, no matter how obliquely, on the weather. Clearly, when we are in the public sphere, we are only minimally social. Nonetheless there is a social component. There is, however, the possibility that ICTs and mobile communication will take a small bite out of the already minimal sociability that is available in this sphere.

Coming back to the more general question posed at the start of this chapter: Can the social consequences of the mobile telephone or even of the transistor be measured against those caused by the printing press, the steam engine, or the automobile? Will my life or, more to the point, that of my children and eventual grandchildren be changed?

More than half a century has passed since the development of the transistor. As noted earlier, the general form of the major institutions seems to be intact. However, as Beck and Beck-Gernsheim point out, these developments may facilitate the rise of a new institution, that of individualism (2002). Indeed, in his introduction to Beck's book, Scott Lash takes up just this interaction between technology and society. In a sense, individualism will reform the other institutions in its own image. The educational system, the family, the political structure, and all the others — their tasks will be carried out against the backdrop of the individual and his or her uniqueness, ego, and particularities. While there will still be altruism, it may be of a type that is more carefully calculated and applied where one can expect the greatest effect. While our sense of reciprocity will live on, it may be a type of rationalized and strictly focused version. The role of the mobile telephone does not necessarily seem to be causal here, but it is a good supporting actor.

I started this book by referring to my grandfather's efforts to develop radio-based telephony in the late 1940s. In this chapter, I began by drawing on Osborne's prognosis as to the eventual range of mobile telephony from about the time of my birth. Perhaps it is best to end with some speculation as to what my children will experience. Will they have a mobile communication device that is ubiquitously Orwellian, a device that allows me to call them whenever or wherever they are? I hope not. While I wish them the ability to contact me — and me them — as needed, they have to develop a life that is beyond my reach. While I hope to be there for them when needed as they mature into teens and then young adults, it will increasingly be their prerogative to define those situations.

Will the mobile telephone enable them to coordinate their activities, feel secure, and participate fully in their various social circles? I dearly hope so. Will their eventual use of mobile communication become a social crutch, or close them off to others? Will they exist in a type of virtual walled community in which they interact only with those whom they deem of interest at the moment? Just as dearly I hope not. To live in such a world is to lose our empathy and our ability to experience the different phases of life.

Will they have a mobile communication device that allows them to share their delights and their frustrations with me? I hope so. It is here that we come to the most central issue. I hope that the future of mobile communication will allow us to develop and share common experiences and insights. I hope that it will allow us a richness that will keep our relations intact between meetings, and I hope that it will fit so well into our lives that it will not hinder our participation in the here and now.

APPENDIX

Data Sources Used in the Analysis of Mobile Telephony

Name of the analysis	Year	Type of analysis	Country	General description
Telenor in-home interviews	1997–2000	In-home interviews	Norway	This series of studies consists of two rounds of interviews carried out in 1997 and 2001, respectively. These interviews generally focused on examining how families with teens and children reacted to the growing role of ICTs in everyday life. There was also an emphasis on the planning associated with ICTs and their impact on the routines in the home. The material from the interviews was transcribed and then examined via usage content analysis techniques. The interview rounds included 10–15 homes per round.
Telenor observational studies	2001–2002	Observational	Norway	This study is an observational study of mobile telephones in natural settings. The observations were made in a variety of public locations and noted *in situ* as field notes. The study was carried out in 2001 and 2002. Over 200 observations were made.

Name of the analysis	Year	Type of analysis	Country	General description
EURESCOM P903 qualitative material	1999	Focus groups	Nine European countries	The qualitative side of the Eurescom P903 study included 36 focus groups in six countries across Europe. Six groups were convened in each country, examining those who did and did not use the mobile telephone and/or the Internet. The material from the focus groups was gathered and examined by a team of social scientists from the respective countries plus other members of the P903 project group.
Telenor focus groups	1995–2003	Focus groups	Norway	Telenor has held several series of focus groups to examine the use of mobile telephony and the more general use of ICTs in everyday life. The groups in the separate series are typically divided into age- and/or gender-based groups. The sessions are recorded and often transcribed for analysis via content analysis techniques. There are four major series of groups in this category. These have taken place in the period between 1995 and 2003 and typically included sessions for persons from different age groups, for persons with different access to ICTs, and, at times, for groups defined by gender.
Telenor surveys	1997–2002	Survey	Norway	Telenor has carried out a series of surveys in Norway with nationally representative samples of individuals annually between 1997 and 2002. The sample size varied between 1 and 2000, depending on the need to oversample various populations. The surveys during the first 3 years of this series focused exclusively on teens and their mobile telephone use. In the later surveys all age groups were included.

EURESCOM P903 quantitative material	2000	Survey	Nine European countries	The EURESCOM P903 study included 9079 face-to-face interviews from a randomly selected sample of persons in The Czech Republic, Denmark, France, Germany, Italy, the Netherlands, Norway, Spain, and the United Kingdom. The data was collected in 2000.
Statistics Norway media use	1994–2002	Survey	Norway	Every year since 1994 Telenor has participated in Statistics Norway's survey of media use. This is an annual survey in which the data is collected on a quarterly basis. The questionnaire includes a broad variety of questions about media use. Typically, the survey queries 1–2000 persons. This is a nationally representative sample.
E-living study	2001–2001	Panel survey	Five European countries	The e-living database is a panel survey that, in the first round, included 10,534 interviews with persons from the United Kingdom, Italy, Germany, Norway, Bulgaria, and Israel. The follow-up data includes interviews in 7205 of the original homes and additional interview material with other household members as well as a time diary.
Pew Internet & American Life Project October 2002 tracking survey	2002	Survey	United States	This is supplemental material.
International Telecomm Union data on mobile ownership	2002	Survey	Worldwide	This is supplemental material.
Rutgers studies of Internet use	2000	Survey	United States	This is supplemental material.

Endnotes

Chapter 1

1. In this book, the terms *mobile telephone* and *mobile telephony* are used as synonymous with *cellular telephone* and *cellular telephony,* the terms more commonly used in the United States. I prefer the former, for a couple of reasons. First, *mobile telephone* is the popular form of reference in many countries. In addition, *cellular telephony* refers to a technology, whereas *mobile telephony* perhaps better reflects the social dimensions of the technology.
2. In addition, there was a second article in the rival *Rocky Mountain News*, the transcript of a radio interview in which my grandfather participated and a copy of an article he coauthored in the journal *Electrical Engineering*.
3. This is not to ignore the effect of the device on the work world. Indeed, it is often via the business use of the device that people become familiar with mobile telephony. However, my data and my analyses have focused on the private sphere and thus I need to be true to them. For an analysis of mobile telephony in the business world, you can examine, for example, the work of Julsrud (2003).
4. Interestingly the development of "push-to-talk" systems are reviving this one-way form of interaction.
5. This was typically boys who had a subscription with each of the network operators and who, in addition to having more mobile telephone handsets to play with, reasoned that it allowed them to save money. This is because calling others who subscribe to the same company is cheaper than calling across network operators.
6. In the GSM system, this is true as long as you call from your home country. If you roam to a second country, you then assume the cost of forwarding a call.

7. The pricing system in the United States favors packages where, for a fixed amount per month, you receive a certain number of relatively low-priced minutes. If you exceed the ceiling, the per-minute price becomes quite high. The price per minute under the ceiling is typically between 5 and 10 cents; if you exceed the ceiling, the price per minute can be as much as 30–45 cents. The minute ceilings are often tiered; in that is, a more restrictive ceiling applies to traditional calling periods while a more liberal ceiling applies to evenings and weekends. This is an effort to flatten out system use. In addition, there are various approaches to long-distance and roaming charges that apply when you're in not in your "base" location.
8. Some satellite-based systems, such as Iridium, promise global coverage. However, their relatively high cost and limited access to terminals have meant that these are peripheral technologies.
9. $F(3, 5280) = 23{,}955$, sig. < 0.001.

Chapter 2

1. For a critique of this perspective see Hutchby (2001).
2. This is obviously the concept taken to its extreme. Objects have a preferred interpretation. That is, they are designed to emphasize one or another interpretation. Thus, a hammer cannot be interpreted as a pillow or vice versa. Nevertheless, these interpretations are situational and can be redefined at will by different users.
3. Information and communication technologies (ICTs) is a shorthand reference to all types of technologies—usually electronic—through which we can gather information or through which we can communicate. In its broadest interpretation it includes televisions, radios, video/DVD players, satellite dish receivers, PCs, etc. A more restricted definition includes only interactive technologies, such as Internet-enabled PCs (with their associated modems or broadband connection equipment) traditional landline telephones, and mobile telephones.
4. There are indeed instances of platinum- and diamond-encrusted mobile telephones.

5. We can also discuss the role of Roger's opinion leaders in this context (1995).
6. These names, as all the names of interviewees, are fictitious.

Chapter 3

1. The emergency number in the United States.
2. This situation was reported by informant "X" on the *Smart Mobs* Web page administered by Howard Rheingold, http://www.smartmobs.com/.
3. Lemish and Cohen note that the phone provides us with "peace of mind" (Lemish and Cohen 2003).
4. For more illustrations see the discussion at http://www.textually.org/textually/archives/001865.htm#001865 (Textually.org 2003).
5. This data comes from the EURESCOM P903 study. See the Appendix for a fuller description of the data.
6. The only two other indicators that had nearly the same score were an item asking if the mobile telephone was useful for calling and announcing that you're running late (63.4% complete agreement) and an item asserting that the mobile telephone was a hazard when driving (62.1%). The former item will be discussed in Chapter 4, and the latter item will be examined later in this chapter.
7. Chi^2 (5) = 27.78, sig. <0.001.
8. Chi^2 (35) = 82.82, sig. <0.001.
9. Use of the mobile telephone in emergency situations is also noted as a benefit of the Grameen system in Bangladesh (TDG 2002).
10. A corollary problem is the issue of false emergency calls. These can take the form of either willful prank calls or inadvertently dialing of the emergency number sequence. Regardless of the way in which these false calls are generated, they potentially tie up resources and personnel when they may be needed for other purposes.
11. The exact definition as to what constitutes an "emergency" is also a dynamic issue (Klamer *et al.* 2000). This can range from car accidents with injuries, to the urgent request for help carrying in the groceries from the car.

12. Interestingly, there is often incomplete knowledge as to when you can actually use a mobile telephone in an emergency. In a group interview with teens, the discussion showed that some of the participants were unsure about how the emergency features of a mobile telephone would actually function.

    Ida, 18: *I have a question about prepaid [pay-as-you-go] cards. If you don't have money in your account, does it cost anything to call, for example, an ambulance or the police?*

    Moderator: *You can always call.*

    Ida: *And it doesn't cost anything? From anywhere?*

    Moderator: *Regardless, from a pay phone, from home, regardless, it is free.*

    Marianne, 17: *If you turn off the [mobile] phone and turn it on again, before you put in the PIN code it says that you can call SOS, at any rate with my phone.*

    The discussion shows that the teens were interested in the issue and, among other things, saw the mobile telephone as a type of lifeline. Emergency access is part of the GSM specification (ETSI 1996). According to that specification, emergency calls supersede the need for PIN codes, subscription status, etc. Thus, the only requirement for making an emergency call from a GSM telephone is that it have a functioning terminal.

13. Parents often provide their adolescent children with mobile phones with the notion of security as at least a partial motivation.

14. In the wake of the disaster, there was some discussion about using the mobile telephone as a type of rescue beacon in order to find victims in the rubble. It must be realized, however, that the mobile telephone is not really designed for this. It is not like the avalanche beacons that cross-country skiers often use to find others who have been caught in an avalanche. Avalanche beacons are intended to be used over short distances, they are designed to be watertight, and it is assumed that they are used in wilderness areas. Most mobile telephones are not designed for those conditions. In addition, if we think of using a signal from a mobile phone in the type of situation that arose in the World Trade Center, there are other considerations. Radio signals are reflected and suppressed by metal and concrete.

This reduces the precision with which we can locate the source of the signal.

15. It can be noted that the same pattern was not evident in the case of the Pentagon, where security issues as well as military culture appear to have been a barrier to this type of interaction (Dutton 2003). There is also the irony that, in some cases, it was the inability to contact family via their mobile devices that people began to learn of their fate.
16. An issue that is often discussed in this context is that of electromagnetic radiation. This will not be discussed here since analysis of the issue is not decisive. This is more of an individual, as opposed to a social, issue and, perhaps most importantly, I have no specific knowledge on the issue.
17. See also Garfinkel for a discussion of social network analysis in the case of covert and guerilla groups (Garfinkel 2003).
18. Chi^2 (4) = 56.81, sig. <0.001.
19. Chi^2 (28) = 113.63, sig. <0.001.
20. Automobile-based sightseeing also suggests that our attention can be "divided" between driving the car and engaging in cultural enrichment. The radio, advanced stereo systems, and now even video-based entertainment (intended for those in the back seat) are other obvious examples of using the time in the car to engage in secondary activities.

Chapter 4

1. The data comes from the Eurescom P903 database (See the Appendix for a more complete description of the data). The questionnaire contained a battery of questions asking how users felt about various dimensions of mobile telephony. Using factor analysis, a "coordination" factor emerged from the data. This included items such as "The mobile phone helps one to coordinate family and social activities," "Using a mobile phone helps one to notify others when you are late," "I want to call people wherever I am," "Having a mobile phone allows one to enjoy more of your leisure time," "A mobile telephone allows one to do several things at the same time," "I should be

reachable anytime, anyplace," and "A mobile telephone helps one to stay in steady contact with family and friends."

2. Chi^2 (24) = 215.8, sig. <0.001.
3. Chi^2 (4) = 12.4, sig. = 0.015.
4. Chi^2 (24) = 274.5, sig. <0.001.
5. Chi^2 (4) = 31.1, sig. <0.001.
6. Indeed the study of logistics, planning, and coordination are central issues in the administration of both formal organizations (Blount and Janicik 2000) and informal organizations (Rakow 1988).
7. Hägerstrand's studies of time and distance show, in an elegant way, the paths of individuals as they go about these activities (1982).
8. As the mobile telephone is adopted by teens, it weakens this already tenuous link. The fact that the telephone allows geographic mobility in addition to the notion that the device can be turned off (or you can assert that the batteries were low) means that children are freer with this technology (Vestby, 1996). In Norway, many teens have adopted mobile telephones, and the trend is accelerating.
9. Some town clocks, such as the astronomical clock known as *Orloj* in the Prague Town Square, became miniature morality plays, with a parade of various biblical actors playing out their parts every hour on the hour.
10. According to Dohrn-van Rossum (1992), certain crimes were associated with the need for torture in order to extract a confession. The length of time a person could be tortured was eventually fixed.
11. Indeed, the first clocks did not have the now-familiar face at all. Even after the face of the clock appeared, there were several alternative arrangements of the hours, e.g., a 24-hour system starting at sunrise and 12-hour systems. It was not until 1400 that there was even general agreement as to the number of hours in a day and their arrangement. In addition, it was not until the mid-1600s that clocks had the now-standard arrangement of hour, minute, and second hands (Andrewes 2002).
12. While one could determine one's north–south position by reference to the stars, nature does not provide an equally convenient method for fixing one's

east–west coordinate. This was made terribly obvious in 1707, when 2000 sailors and officers in four British ships died because of their inability to judge their east–west position. Prompted by this disaster, Parliament established a prize for the person who could solve the so-called longitude problem.

13. Beniger notes, for example, that the wristwatch became more common at the turn of the twentieth century, after it was employed by the British to coordinate troop movements in the Boer War (Beniger 1986, p. 327).

14. Beyond mechanical timekeeping, people use a range of sociocultural devices to coordinate their activities. These include explicit schedules and routines, implicit pace, and sociotemporal norms (Blount and Janicik 2001).

15. I will come back to the issue of mobile telephones and their use in public places in Chapter 6.

16. As noted in the previous chapter, the temptation to do this while driving a car has its negative consequences.

17. Within this context, there are often gendered differences in the degree to which we attend to these types of domestic tasks. Analysis has shown that women often have more complex travel patterns and greater "effort distances" in their daily lives. These effectively reduce their range of action (Hayden 1984; Falk and Abler 1980; Hjorthol 2000). Given this difference, we also see that the telephone allows women a way in which to help organize these tasks (Rakow 1988; Vestby 1996; Rakow 1992; Rakow and Navarro 1993; Moyal 1989). Nonetheless, women, and in particular middle-aged women, have been less likely to adopt mobile telephony (Ling and Haddon 2001).

18. I am indebted here to Leysia Palen for insight into the social dynamics of schedules.

19. This material comes from the 2001 Telenor study of a nationally representative sample of 2000 Norwegians. Respondents were asked to read (with exact orthography, grammar, and punctuation) the most recent SMS messages they had sent.

20. The EURESCOM P903 quantitative data.

21. $F(6,8790) = 17.742$, sig. > 0.001

22. $F(1{,}8808) = 0.162$, sig. > 0.686
23. Though I can imagine that the more imperative "IM ON THE BUS NOW" might be in response to the pestering of an anxious parent wanting to know when her child is getting home.
24. While the mobile telephone has this potential, one can also assert that it also encourages factionalism, which is anathema to the success of protest movements (Ling 2000b; Gamson 1975).
25. As will be discussed in Chapter 8, swarming via the mobile telephone is successful to the degree that there was a common focus for the actions, a clearly understood and accepted strategy for the action, and no fractionalization in the movement at either the ideological or the strategic level.
26. This data comes from the EU e-living project.
27. Based on the data gathered in Norway in May of 2002.
28. Another issue here is that in many cases the mobile telephone is not as securely fastened to your body as your watch. This offers a further hazard in that the device can fall out of your pocket and be dashed against the floor. In Korea, this has been addressed by attaching the mobile telephone to a neck strap. Thus, the device is firmly attached to the body while at the same time being accessible for use.
29. The one exception seems to be the use of 16:02 as a time to meet. We can perhaps interpret this as an absurdly precise part of a humorous exchange.

Chapter 5

1. This material comes from the annual Statistics Norway media use study.
2. "A" is the person making the call, and "AM" is the person who answers the telephone. These are the names provided in the transcript reported by Veach.
3. Interproxminate interaction describes the management of the physical space between individuals as they interact. Interkinesic interaction the dialogue of

gestures in social interaction; i.e., shrugging our shoulders at the appropriate moment, nodding, the use of hand gestures, etc.

4. The staging of these varies from country to country in terms of how you introduces yourself and the information that you offer at which stage in the opening (Holmes 1981).
5. See also Bjelić (1987) and Mininni (1985) for a discussion of children's mastery of linguistic skills over the telephone.
6. It is quite possible that some aspects of intonation are an issue related to the development and mastery of motor skills.
7. In Norway, children of divorced parents are sometimes equipped with a mobile phone in order to facilitate access between the nonresident parent and the child.
8. In Norway, the vast majority of individuals are members of the state church, and thus confirmation is a common experience for most teens. Indeed, there is a civil confirmation for those who are not members of the state church. It usually occurs when the child is 14 or 15 years old.
9. I have discussed safety, coordination, and emancipation in the same breath when looking into the needs of the wheelchair bound and the hearing impaired.
10. Interestingly, there are gender differences between parents when considering the acceptability of using mobile telephony for the resolution of everyday coordination issues. Earlier research has shown that fathers are more likely to accept the use of technology to solve such coordination issues (Ling 1999).
11. Manceron describes some of the same in her analysis of Parisian adolescents (1997).
12. A Philips Fizz is a reference to an earlier, relatively unpopular mobile telephone. These comments come from a series of focus groups for teens and young adults held by Telenor in 2003.
13. Our audible presentation is also dictated by style and situation. The use of "r" by New Yorkers from various class backgrounds shows how speech is an element in the presentation of self that varies with fashion and circumstances (Bryson 1991, p. 101).

14. See also, for example, Forsythe *et al.* (1985), Harp *et al.* (1985), and Rucker *et al.* (1985).

15. Simmel has rightly been critiqued as being too class based in his analysis of fashion (Davis 1992; Sproles 1985). Others have looked into the influences of various social groups on the introduction of styles (Polhemus 1994).

16. It can also help those who act on stereotypes to select those who are "legitimate" objects for retribution, as in the case of the so-called Zoot Suit riots during the Second World War (Cosgrove 1984; Mazon 1984; Polhemus 1994). By adopting a style of dress or display or even a style of language or form of courtesy, they can show their sympathy or antipathy for a group (Duncan 1970, p. 269). While display is generally associated with the idea of fashion, we can also examine language use in these terms. In this case, there is an equally rich and dynamic form of interaction, particularly among youth (Fine 1987).

17. $F(7, 1686) = 90.531$, sig. <0.001.

18. $F(1, 1692) = 38.436$, sig. <0.001.

19. It is interesting to note that it is also among the teens that the largest number of names are in active use. While this is only a small portion of the total number of names recorded in the telephone, an examination of this material shows that it is the younger users who are the most active socially. For the whole sample, only about two are used on a daily basis and six on a weekly basis. For teens and young adults, about 10 are called on a weekly basis and three to four on a daily basis. We can see in these numbers that among the younger users there is a much broader arena of social contact.

20. While adults often express the sense that the mobile telephone results in stress (Eriksen 2001), teens have the opposite perspective (Stuedahl 1999).

21. Data from the EURESCOM P903 database.

22. Pearson $r = 0.56$. All correlations noted here are significant at more than the 0.001 level of significance.

23. Pearson $r = 0.30$.

24. Pearson $r = 0.31$.

25. Pearson $r = 0.56$.

26. Pearson $r = 0.44$.

27. Pearson $r = 0.42$.

28. Pearson $r = 0.49$.

29. $F(3, 4335) = 13.45$, sig. <0.001.

30. $F(3, 4476) = 12.444$, sig. <0.001.

31. $F(3, 4422) = 3.545$, sig. $= 0.014$.

32. $F(1, 11039) = 81.03$, sig. <0.001.

33. $F(3, 4242) = 24.203$, sig. <0.001.

34. $F(3, 4306) = 3.830$, sig. $= 0.009$.

35. Prepaid, or "pay-as-you-go," subscriptions have no fixed cost associated with them, and so the cost per call is generally somewhat higher than for postpaid subscriptions. In the context of mobile telephony, operators have developed a system whereby you can buy additional access electronically, thus simplifying the administration. You simply buy a card containing an access number that, when entered into the telephone, increases your account balance by a corresponding amount. In this system you pay beforehand for telephone access. Thus, when the allotted time or sum of money has been consumed, you cannot call out to other (nonemergency) numbers. You can, however, continue to receive calls and text messages for a certain amount of time, which varies by network operator.

36. In this system, stores selling terminals receive a rebate from network operators for the sale of specific types of terminals if a subscription is included in the sale. The person purchasing the terminal/subscription often binds himself or herself to the network operator for a certain period. From the position of the network operator, this is an effective way to encourage the use of that operator's network.

37. The material in this analysis comes from the 2002 survey of Norwegian mobile telephone use. This included a random sample of 2002 Norwegians over 13 years of age.

38. The material here shows that 78% of the men and 91% of the women pay for mobile telephony from their personal income. Thus 22% of the men

have their mobile telephone use subsidized by their employers. This is a statistically significant difference: chi^2 (1) = 52.857, sig. <0.001.

39. Data shows that one in four 16-year-olds has a job, whereas over three of four 19-year-olds have a job (Vaage 1998). Analysis has also shown that working teens are more likely than others to own a mobile telephone (Ling 1998a). Analysis indicates that in 1995 Norwegian teens used about 1575 Nkr (ca. $175) per month for the purchase of various items (Brusdal 1995). This material was gathered before the widespread use of the mobile telephone by teens. A Danish study found that 16-year-olds have a mean of 1700 Dkr. (ca. $200) spending money per month (Dortner 2000).

40. Using a slightly different metric, we find that young adults spend about 2% of their total household income—not disposable income as in the previous analysis—on mobile telephony. Older adults reported spending about 1% of their total household income for personal mobile telephone use. Obviously, this measurement is colored by the fact that household income can be the aggregate of several individual incomes.

41. It is more expensive to call from a landline phone to a mobile phone than to call from a landline phone to another landline phone.

42. For their part, parents can also adopt various guerrilla tactics when trying to call their child. If, for example, they are unable to reach their child, they can begin to call others in their child's social network. In this way they can eventually tease out information as to where their own child is and perhaps even reach the child directly via the friend's mobile phone.

Chapter 6

1. Chi^2 (12) = 70.93, sig. <0.001.

2. Singles bars and restaurants where one eats in order "to be seen" are in a class by themselves. In the singles bar, the barriers are there to be opened and closed as the attentions of others are sought or to be avoided (Collas 1995; Giuffre and Williams 1994; Haavio-Mannila and Snicker 1980; Parker 1988). In the case of restaurants where we hope "to be seen," we may even use strategies to underscore our presence. Strategies include the grand entrance

and sending a message to the management to page ourselves over the PA system. This latter strategy seems somewhat similar to the patron who uses the mobile telephone only as a way to draw attention to him or herself.

3. Many of these issues are avoided in the case of text messages, a very popular form of interaction in Europe and Asia. These are different from voice telephony, in that there is no real audio component and they are also asynchronous. Thus, we need not attend to them here and now. When we do send a message, there is no need to cover over the sound.

4. Nonverbal communication has been a popular area of academic focus (Burgoon 1985; Leffler *et al.* 1982; Archer and Akert 1977; Cherulink *et al.* 1978; Ekman and Friesen 1969; Felipe and Sommer 1966; Greenbaum and Rosenfeld 1978; Hall 1973; Quek *et al.* 2000; LaFrance and Mayo 1978; Watson and Graves 1966). Many analyses have looked into the undeniable contribution of gesture, kinesics, proximics, vocalics, appearance, haptics (touch), chronemics, and manipulation of artifacts associated with communication.

5. Partners who are intimate can guide each other by placing a hand on the others back, hips or other body regions that are generally not accessible to non-intimates. Again, there are clear cultural differences in how this is practiced.

6. In this connection the adoption of hands-free devices can complicate this situation, since there are fewer immediate signs. Via their use, we appear to be available for interaction, whereas the opposite is actually the case (Love 2003).

7. In an interesting twist on this, muggers in the United Kingdom have played on the strategy of mobile telephonists being in less populated areas of the subway. Charles Dunstone, the CEO of Carphone Warehouse, suggested that users should stand with their backs to walls when making calls in public to avoid being attacked from behind (Burrell 2002).

8. There are many variations of being embarrassed for others. Such forced eavesdropping is only one version of this. I am reminded of stories (or perhaps actual experiences) of men going into important meetings with their fly down or women having menstrual blood showing in the crotch of their pants that can arouse the same sense.

Chapter 7

1. In the United States there are two-way paging systems, such as Blackberry, which have some similarities to the texting systems described here.
2. In the United States, there are several reasons for the slow adoption of text messages. Carriers have been slow to to cooperate in the transmission of text messages across company boundaries. In addition, many handsets in the United States do not have the functionality required to produce the messages. The relatively high use of the Internet and the use of instant messaging has provided some of the same functionality as mobile text messaging. Finally, there are issues of coverage that have augured against the adoption of SMS.
3. This assertion is based on an analysis of data from the EU's e-living project from 2001.
4. Chi^2 (4) = 15.85, sig. = 0.003.
5. Chi^2 (28) = 793.72, sig. <0.001.
6. According to Hashimoto, this picture is quite similar to that in Japan. He reports that 96% of women 20–24 years old are mobile telephone users. "Only" 81% of the men in the same age group are mobile phone users (Hashiomoto 2002). A somewhat similar use pattern is found in the work of Roessler and Hoeflich (Roessler and Hoeflich 2002).
7. $F(1,1775) = 9.58$, sig. = 0.001.
8. $F(7,1769) = 72.89$, sig. < 0.001.
9. It also shows that if we look at older teens, there is a transition over to mobile voice use.
10. Originally, SMS was a free service. In Norway, it gained a sudden popularity in the summer of 1998, when SMS traffic grew to the verge of overwhelming the system of voice traffic. At this point tariffs were introduced. Currently an SMS message costs somewhere between 10¢ and 15¢. By contrast, a standard telephone call from a mobile telephone with a prepaid, or "pay-as-you-go," subscription costs between 25¢ and 90¢ per minute, depending on the time of day and the type of telephone being called.

11. A common version is the so-called "T9" – or Text on 9 keys—method of text entry. It is a commercial product that is basically a dictionary linked to the appropriate keys so that instead of doing multiple taps to get to a letter, you more or less spell out the word and the telephone suggests a word. Most of the time it is the correct word, particularly with longer words, since it can narrow down the universe more. The program does not add automatic abbreviations unless you code them in. You actually have to tap in all the letters of a word in order to enter it. More than anything else, it adds to the speed of the writing, not necessarily the length of the messages. Various programs attempt to extend this to word and even phrase prediction.
12. The material described here was gathered by Telenor in 2002 among a representative population of Norwegians.
13. Mobile photography further enhances our ability to inform others of our situation.
14. Yet another form of integration provided by texting is seen in the various forms for SMS gaming (Rheingold 2002).
15. Indeed, material from the *Ung I Norge* study shows a disturbing correlation between intense SMS and voice mobile use and other forms of deviant activity. The material in the study shows that there is a covariance between mobile telephone use (both voice and SMS) and various boundary testing behaviors (such as the degree of sexual experience). In addition, the material indicates that there is an interaction between mobile telephone use and participation in illicit behaviors for a small portion of teens (Ling forthcoming).
16. The analysis in this section is based on a corpus of SMS messages gathered in a Telenor survey from a random sample of just over 2000 Norwegians in May of 2002 by telephone. Along with demographic, behavioral, and attitudinal questions associated with mobile phone and SMS use, we asked the respondents to read (and, where necessary, to spell out) the content of the last three messages they had sent. This resulted in a body of 882 SMS messages from 463 (23%) of the 2002 respondents. The data show that 64.2% of the sample reported sending an SMS at least once a week. Thus, a significant number of SMS users did not provide any messages. Not all respondents agreed to read their messages, there may have been selective filtering of content since respondents may not have wished to particularly revealing

or piquent messages and not all respondents had saved three messages. There are no statistically significant age or gender differences between those who did or did not provide messages.

The approach has several advantages, most notabily that it is one of the few studies that strives to have messages from all age, gender, and socio-demographic levels of society. However, we need to be aware that the method may have colored the data. There is an ethical and a methodological reason we asked for the last messages sent as opposed to those received. Ethically, it is not possible for the researcher to ask for messages a respondent has received, since implicitly we would include data from persons who had not given their consent to participate in the study. Further, we do not know the background, demography, or other characteristics of the sender for messages a respondent has received. Thus, it is not possible to analyze the material in any meaningful way at a sociological level.

The messages were read by the respondent to the interviewer, who then transcribed them into the database. Given the tendency to use both intended and unintended abbreviations and misspellings in the messages, we can suspect that the transcription process resulted in some errata, emendation, and morphing. A limited point of control is offered via a study of SMS texts generated by 82 teens from Grimstad, a small town in southern Norway (Ling and Sollund 2002). In this case, the respondents filled out a questionnaire as opposed to being interviewed. Thus, they were asked to transcribe SMS messages themselves. While it is difficult to quantify the differences, given the smaller size of the Grimstad sample, the reader is left with the impression that this more direct form of data collection resulted in a slightly "rougher" corpus. The teens seemed willing to include somewhat more profanity and unguarded remarks.

Another weakness with the more general Telenor material is that the messages are often take out of their context in a sequence of messages sent to another person. Obviously, this can make interpretation difficult in some respects, and it eliminates the possibility of doing any type of discourse analysis. Unfortunately, this limitation comes with the territory. As already noted, to do any type of data collection for which one does not have the consent of the authors of the text is not ethically defensible.

In spite of these various caveats, the sample of messages reflects SMS use across age, gender, and socioeconomic groups. We gain insight into the phenomena, and the material allows us to generalize the results to a greater degree than in convenience samples.

17. A somewhat similar analysis of mobile voice telephony shows that people most often use the device for instrumental communication. In a 2001 survey carried out by Telenor, approximately 60% of all calls were reported to be for concrete purposes, such as coordinating activities and exchanging information. About one-third of all calls were reported to be focused on more expressive social interaction. The remaining 5% were reported to be "humorous" calls. It is perhaps not surprising to note that men, in particular middle-aged men, dominated the "instrumental" calling category. Young adult women dominated the expressive calling category, and young teen boys dominated the "humorous" category. The work by the Finnish researcher Arminen points to the various ways in which location—and thus coordination—are important in voice mobile communication (2003).

18. Chi^2 (1) = 3.35, sig. = 0.067.

19. Chi^2 (1) = 4.76, sig. = 0.029. Planning in the middle-range future was defined as making agreements for activities that had not already started and were to take place within the next few days.

20. Chi^2 (7) = 17.83, sig. = 0.013.

21. Chi^2 (1) = 4.77, sig. = 0.029. Planning in the immediate future was defined as messages involving planning for activities that were already in progress such as "*I am on my way home*" (M, 38) or "*IM ON THE BUS NOW*" (F, 15).

22. Chi^2 (1) = 8.77, sig. = 0.003.

23. Chi^2 (7) = 13.28, sig. = 0.066.

24. Chi^2 (1) = 9.634, sig. = 0.002. Emotionally based grooming messages were typically greetings that included declarations of love.

25. Indeed, Sadie Plant, in her international analysis, suggests that "Most text messages sent and stored by teenagers from Birmingham to Bangkok have some relation to sex" (Plant not dated, 80). To put it in somewhat more sober tones, the material here points in other directions.

26. Grooming messages consisted of those messages that did not have an obvious instrumental function, but rather seemed to indicate a less-directed form of interaction. These messages seem to describe the nurturing of

friendships and romances. Taylor's notion of *gifting* can perhaps be applied most directly to this type of message (Taylor and Harper forthcoming).

27. In the SMS messages shown here are loyal to the original material, in that in translating, I have tried to retain the "spelling conventions", punctuation, capitalization, and flavor of the original Norwegian texts.

28. This analysis was done with the aid of the software provided by Yomans and available at http://www.missouri.edu/~youmansc/.

29. In a sample of spoken Norwegian, though a poor one, the article *det* (it), the pronoun *jeg* (I), and the verb *er* (are) are the most frequently used words for both men and women. The complementary pronoun *du* (you) is scarcely seen among the top 10 most used words. The adverb *så* (so) is ranked as the sixth (men)/seventh (women) most used word; *så* (so) makes up approximately 2% of the words in the speech sample. This corpus comes from the transcription of a Norwegian language focus group and is not really up to snuff as a basis for comparison, since it is not natural language but rather talk that is focused on a particular topic. Ideally, we would use telephone conversations, but gaining access to such material is carefully governed in the privacy laws.

30. "Hug" for example was the 38th most used word for women (often used as a closing).

31. The specific measure used here is the type/token analysis. The tool used to carry out this analysis was that developed by Youmans and is available at http://www.missouri.edu/~youmansc/.

32. $F(1,478) = 10.445$, sig. = 0.001. Hård af Segerstaad reports longer messages in Swedish SMS messages. In her analysis, the messages were more than twice as long, at an average of 14.77 words per message. She does not break the analysis down by gender (Hård af Segerstaad 2003b).

33. Chi^2 (1) = 9.87, sig. = 0.001.

34. Chi^2 (1) = 9.64, sig. = 0.001.

35. Hård af Segerstaad reports even more modest use of abbreviations in the work she reviewed (Hård af Segerstaad 2003a).

36. In the case of the study carried out in Grimstad, the number of abbreviations was about 10% higher, though this was exclusively a teen sample. This leads us to suspect that some of the abbreviations did not survive the data-collection process. In terms of the demographic analysis presented here, it is hoped that the bias imposed by the transcription process was similar across all the messages.
37. Chi^2 (1) = 35.19, sig. <0.001.
38. Chi^2 (1) = 9.30, sig. = 0.002.
39. Chi^2 (1) = 4.17, sig. = 0.002, and Chi^2 (1) = 3.41, sig. = 0.06, respectively.
40. Ellwood-Clayton describes the stylized use of punctuation in her work in the Philippines (Ellwood-Clayton 2003).
41. Chi^2 (1) = 7.35, sig. = 0.007.
42. Chi^2 (7) = 21.33, sig. = 0.003.
43. Chi^2 (7) = 14.99, sig. = 0.036.
44. Chi^2 (7) = 25.87, sig. < 0.001.
45. Chi^2 (1) = 4.98, sig. = 0.025.
46. Chi^2 (7) = 17.48, sig. = 0.014.
47. The classic example is a message to the effect of "It is not just because I am drunk that I say that I love you."
48. In most other countries, the general rule is that the calling party pays, except when they have "roamed" outside of the particular country in which the caller's subscription is based. At that point, the receiving party pays to have the call forwarded.

Chapter 8

1. The word *motel*, a contraction of "motor hotel," was first used by James Vail to describe his establishment in San Luis Obispo, California, in 1925 (Flink 2001).
2. In a panel design the same persons are interviewed at various points over time. The most normal version of a panel study is simply reinterviewing the

persons once. In versions that are more elaborate, the researcher follows the same persons over longer periods of time.

3. I must underscore the word *maintain* here. The Internet is more open in this context since we can, for example, join chat groups and other relatively open forums for interaction where we meet others who happen to be there at the time. The mobile telephone, by contrast, requires that we have the telephone number of our intended interlocutor. This means that in the vast majority of cases we have at least a passing relationship to our interlocutor before we have a telephonic relationship. The mobile telephone allows us, in many cases, to limit access (Fortunati 2002). Indeed, mobile telephone numbers are rarely provided in telephone directories, and the users of prepaid, or "pay-as-you-go", subscriptions are, in some cases, completely anonymous. While these may be easy to come by (indeed teens seem to hand theirs out at the drop of a hat), the structure of the system is still slightly more closed than we find in the Internet world.

4. Indeed, in the United States, the pricing system that allows for "nationwide unlimited weekend minutes" means that you can call friends, family, and even acquaintances in other states and cities for the most trivial reasons.

5. The experience of the United States and Europe seems to differ slightly here. The "per-minute" pricing of mobile telephony in Europe means that it is often used for the organization of the local group and rarely for intercity or international calls. By contrast, nationwide pricing plans in the United States give one relatively easy access to those who live far away. The pricing system in the United States might have a different effect on the way the device is used. The "nationwide" anytime minutes might allow one to maintain formerly strong ties over greater distance. One can call friends and relatives across great distances for relatively minor things. Thus, the ties are maintained and perhaps even strengthened to some degree. However, this remains within the idea of bounded intimacy, in that one's social horizon is maintained, perhaps at the expense of newer additions. Also, one can wonder whether the intimacy can be maintained over time. It would probably need more direct rejuvenation from time to time. The pricing plans in the United States are, however, a transitory system that will likely disappear as the "weekend slumps" in system traffic disappear.

6. Interestingly, the device also relaxes our need to know where others physically are, since we know that we can easily summon their attention as needed. I am indebted to Leysia Palen for this insight.
7. This is the core of what Rheingold has called the *smart mob* phenomenon (Rheingold 2002).

Bibliography

Anderson, B., and Tracey, K. 2001. "Digital living: The impact (or otherwise) of the Internet on everyday life."*American Behavioral Scientist* 45(3):456–475.

Andrewes, W.J.H. 2002. "A Chronicle of timekeeping." *Scientific American* 287(3):58–67.

Archer, D., and Akert, R.M. 1977. "Words and everything else: Verbal and nonverbal cues in social interpretation." *Journal of Personality and Social Psychology* 35(6):443–449.

Arminen, I. 2003. "Location: a socially dynamic property — A study of location telling in mobile phone calls."In *The Good, the Bad and the Irrelivant: The Users and the Future of Information and Communication Technologies*, edited by Haddon, L., *et al.* Helsinki: Media Lab.

Aronson, S.J. 1971. "The sociology of the telephone." *International Journal of Comparative Sociology* 12(3):153–156.

Arvind S., Svenkerud, P. and Flydal, E. 2002. "Multiple Bottom Lines: Telenor's Mobile Telephony Operations in Bangladesh." *Telektronikk*, 98(1):153–160.

Aveni, A. 2000. *Empires of Time: Calendars, Clocks and Cultures.* London: Tarus Parke.

Bakke, J.W. 1996. "Competition in mobile telephony." *Telektronikk* 96(1):83–88.

Bakken, F. 2002. "Telenærhet: en studie av unge døve og hørendes bruk av SMS." *Sosiologi*. Oslo: Universitetet i Oslo.

Baron, N. 1998. "Letters by phone or speech by other means: The linguistics of email." *Language and Communication* 18:133–170.

Baron, N. 2000. *Alphabet to Email: How Written English Evolved and Where it's Heading.* London: Routledge and Kegan Paul.

Baron, N. 2001. "Why email looks like speech: Proofreading, pedagogy, and public face." In *Language, the Media, and International Communication.* Oxford, UK: St. Catherine's College Press.

Baym, N.K. 2002. "Interpersonal life online." In *The Handbook of New Media*, edited by Lievrouw, L., and Livingstone, S. London: Sage.

Bech, T. 2002. *Skikk og Brukk: Vett og uvett.* Oslo: Kagge forlag.

Beck, U., and Beck-Gernsheim, E. 2002. *Individualization: Institutionalized Individualism and Its Social and Political Consequences.* London: Sage.

Bell, D. 1980. *The Winding Passage: Essays and Sociological Journeys 1960–1980.* New York: Basic Books.

Beniger, J.R. 1986. *The Control Revolution: Technological and Economic Origins of the Information Society.* Cambridge, MA: Harvard University Press.

Berger, P., and Kellner, H. 1964. "Marriage and the construction of reality." *Diogenes* 45:1–25.

Berger, P., and Luckmann, T. 1967. *The Social Construction of Reality: A Treatise in the Sociology of Knowledge.* New York: Anchor Books.

Bijker, W.E., and Law, J. 1992. "General introduction" in *Shaping technology/Building society: Studies in sociotechnical change*, edited by Bijker, W.E. and Law, J. Cambridge, MA: MIT Press.

Bijker, W.E., *et al.* 1987. *The social construction of technological systems: New directions in the sociology and technology of history.* Cambridge: MIT Press.

Bjelić, D. 1987. "On hanging up in a telephone conversation." *Semiotica* 67(3/4):195–210.

Blaise, C. 2000. *Time Lord: Sir Sanford Fleming and the Creation of Standard Time.* New York: Vintage Books.

Blount, S., and Janicik, G.A. 2001. "When plans change: Examining how people evaluate timing changes in work organizations." *Academy of Management Review* 26:566–585.

Boulding, K. 1980. *Beasts, Ballads and Bouldingisms: A Collection of Writings by Kenneth E. Boulding.* New Brunswik, NJ: Transaction Press.

Bourdieu, P. 1985. "The forms of capital." In *Handbook of Theory and Research for the Sociology of Education*, edited by Richardson, J. G. New York: Greenwood Books, pp. 241–258.

Brittain, C.V. 1963. "Adolescent choices and parent–peer cross-pressure." *American Sociological Review* 28:385–391.

Brooks, J. 1976. *Telephone: The First Hundred Years.* New York: Harper and Row.

Brown, B.B. 1990. "Peer groups and peer cultures." In *At the Threshold*, edited by Feldman, S.S., and Elliott, G.R. Cambridge, MA: Harvard University Press, pp. 171–196.

Browowski, J. 1973. *The Ascent of Man.* Boston: Little, Brown.

Brusdal, R. 1995. Rapport 9-1995: *Ungdoms eget forbruk En empirisk studie av ungdommer i alderen 14 til 21 år.* Oslo: SIFO.

Bryson, B. 1991. *The Mother Tongue.* London: Penguin Books.

Burgoon, J. (Ed.). 1985. *Nonverbal Signals.* Beverly Hills, CA: Sage.

Burns, P.C., *et al.* 2002. "How dangerous is driving with a mobile phone? Benchmarking the impairment of alcohol." Crowthorne, UK: Transport research laboratory.

Burrell, I. 2002. "Don't turn your back, says boss of mobile phone firm." pp. 1 in *The Independent.* London.

Bush, D., and Simmons, R.G. 1981. "Socialization processes over the life course." In *Social Psychology: Sociological Perspectives*, edited by Rosenberg, M., and Turner, R.H. New York: Basic Books.

Cahill, S.E. 1990. "Childhood and public life: reaffirming biographical divisions." *Social Problems* 37, 3:263–270.

Cain, A., and Burris, M. 1999. "Investigation of the use of mobile phones while driving." http://www.cutr.eng.usf.edu/its/mobile_phone_text.htm. 5 March 2003.

Castelain-Meunier, C. 1997. "The paternal cord: Telephone relationships between 'non-custodian' fathers and their children." *Reseaux* 5(6):161–176.

Chapman, S., and Schofield, W.N. 1998. "Lifesavers and cellular samaritans: Emergency use of cellular (mobile) phones in Australia." *Accident Analysis and Prevention* 30(8):15–19.

Chen, V. 1990–91. "Mien Tze at the Chinese dinner table: A study of the interactional accomplishment of face." *Research on Language and Social Interaction* 24:109–140.

Cherulink, P.D., *et al.* 1978. "Social skill and visual interaction." *The Journal of Social Psychology* 104:263–270.

Claisse, G., and Rowe, F. 1987. "The telephone in question: Questions on communication." *Computer Networks and ISDN systems* 14:207–219.

Clark, H., and Brennan, S. 1991. "Grounding in communication." In *Perspectives on Socially Shared Cognition*, edited by Levine, J.M., and Teasley, S.D. Washington, DC: American Psychological Association.

Clark, H., and Marshall, C.R. 1981. "Definite reference and mutual knowledge." In *Elements of Discourse Understanding*, edited by Joshi, A.K., *et al.* Cambridge, UK: Cambridge University Press, pp. 10–63.

Clark, H., and Schaeffer, E.W. 1981. "Contributing to discourse." *Cognitive Science* 13:259–295.

Cohen, K., and Wakeford, N. 2003. "Making of mobility, making of the self." Surrey, UK: University of Surrey.

Coleman, J. 1988. "Social capital in the creation of human capital." *American Journal of Sociology* 94:95–120.

Collas, S.F. 1995. "Some of my best friends are ...: Constructing borders and boundaries between self and other, us and them." In *Meeting of the American Sociological Association.*

Collins, R. 1994. *Four sociological traditions.* New York: Oxford University Press.

ComCare. 2002. "Wireless." http://www.comcare.org/research/topics/wireless.html. 3 July 2003.

Conly, R.L. 1954. "New miracles of the telephone age." *National Geographic* July:87–119.

Cooper, G., *et al.* 2002. "Mobile society? Technology, distance, and presence." In *Virtual Society? Technology, Cyberbole, Reality*, edited by Woolgar, S. Oxford, UK: Oxford University Press.

Cosgrove, S. 1984. "The zoot-suit and style warfare." *History Workshop* 18.

Cottrell, W.F. 1945. "Death by dieselization: A case study in the reaction to technological change." *American Sociological Review* 16:63–75.

Crabtree, J., *et al.* 2002. "Reality IT—Technology and everyday life." 2002. London: i Society.

Crawford, M. 1994. "The world in a shopping mall." In *Variations on a Theme Park: The End of Public Space*, edited by Sorkin, M. New York: Hill and Wang, pp. 3–30.

Cunningham, P.A., and Lab, S.V. 1991. "Understanding dress and popular culture." In *Dress and Popular Culture*, edited by Cunningham, P.A., and Lab, S.V. Bowling Green, OH: Bowling Green State University Popular Press, pp. 5–20.

Davis, F. 1985. "Clothing and fashion as communication." In *The Psychology of Fashion*, edited by Solomon, M.R. Lexington, KY: D.C. Heath, pp. 15–27.

Davis, F. 1992. *Fashion, Culture and Identity.* Chicago: University of Chicago Press.

de Gournay, C. (Ed.). 2002. *Pretense of Intimacy in France.* Cambridge, UK: University of Cambridge Press.

de Gournay, C., and Smoreda, Z. 2003. "Communication technology and sociability: Between local ties and 'global ghetto?' In *Machines That Become Us: The Social Context of Personal Communication Technology*, edited by Katz, J.E. New Brunswick, NJ: Transaction Books, pp. 57–70.

de Sola Pool, I. 1977. "The communications/transportation tradeoff." *Policy Studies Journal* 6:74–83.

de Sola Pool, I. 1980. "Communications technology and land use." *Annals of the American Academy of Political and Social Science*, 451(Sept):1–12.

Dichter, E. 1985. "Why we dress the way we do." In *The Psychology of Fashion*, edited by Solomon, M.R. Lexington, KY: D.C. Heath, pp. 29–37.

Dobsen, K.S. 2003. "How Detroit police reinvented the wheel." http://www.detnews.com/history/police/police.htm. 22 Feb. 2003.

Doering, N. 2002. "'Have you finished work yet?:' Communicative functions of text messages." http://www.receiver.vodafone.com/06/articles/inner05-2.html. 16 Jan. 2002.

Dohrn-van Rossen, G. 1992. *History of the Hour: Clocks and Modern Temporal Orders.* Chicago: University of Chicago Press.

Dortner K. 2000. "Difference and diversity: Trends in young Dane's media use." *Media Culture and Society* 22:149–166.

Douglas, M., and Isherwood, B. 1979. *The World of Goods: Towards an Anthropology of Consumption of Goods.* London: Routledge and Kegan Paul.

Duncan, H.D. 1970. *Communication and the social order*. London: Oxford.

Duncan, S. 1972. "Some signals and rules for taking speaking turns in conversations." *Journal of Personality and Social Psychology* 23(2):238–292.

Durkheim, E. 1954. *The elementary forms of religious life*. Glencoe, IL: Free Press.

Dutton, W. 2003. "The social dynamics of wireless on September 11: Reconfiguring access." In *Crisis Communication*, edited by Noll, A.M. Lanham, MD.: Rowman and Littlefield.

Dyckman, J.W. 1973. "Transportation in the cities." In *Cities, Their Origin, Growth and Human Impact*, edited by Davis, K. San Francisco: Freeman, pp. 195–206.

Eisenstein, E. 1979. *The Printing Press as an Agent of Change: Communications and Cultural Transformations in Early-Modern Europe* (Volumes I and II). Cambridge, UK: Cambridge University Press.

Ekman, P., and Friesen, W.V. 1969. "The repertoire of nonverbal behavior: Categories, origins, usage and coding." *Semiotica* 1(1):49–98.

Ellwood-Clayton, Bella. 2003. "Virtual strangers: Young love and texting in the Filipino archipelago of cyberspace" In *Mobile Democracy: Essays on Society, Self and Politics*, edited by Nyiri, K. Vienna: Passagen Verlag, pp. 35–45.

Encyclopedia Britannica. 2002. "Radiotelephone." In *Encyclopedia Britannica 2002: Expanded edition DVD*.

Eriksen, T.B. 1999. *Tidens historie*. Oslo: J. M. Stenersens forlag.

Eriksen, T.H. 2001. *Øyeblikkets tyranni: rask og langsom tid i informasjonssamfunnet*. Oslo: Aschehoug.

ETSI. 1996. "GSM Technical Specification: GSM 02.03 Version 5.0.0." Sophia Antipolis, France: ETSI.

Falk, T., and Abler, R. 1980. "Intercommunications, distance and geographical theory." *Geografiska annaler* 62B(2):59–67.

Farley, T. 2003. "Mobile telephone history." http://www.privateline.com/PCS/history.htm. 22 Feb. 2003.

Felipe, N.J., and Sommer, R. 1966. "Invasions of personal space." *Social Problems* 14(3):206–214.

Fine, G.A. 1981. "Friends, impression management, and preadolescent behavior." In *The Development of Children's Friendships*, edited by Asher, S.R., and Gottman, J.M. Cambridge, UK: Cambridge University Press.

Fine, G.A. 1987. *With the Boys: Little League Baseball and Preadolescent Culture*. Chicago: University of Chicago Press.

Fine, G.A. 1995. "Wittgenstein's kitchen: Sharing meaning in restaurant work." *Theory and Society* 24:245–269.

Fischer, C. 1992. *America Calling: A Social History of the Telephone to 1940*. Berkeley, CA: University of California Press.

Fishman, J.A. 1978. "Interaction: The work women do." *Social Problems* 25:397–406.

Flink, J.J. 2001. *The Automobile Age*. Cambridge, MA: MIT Press.

Flugel, J.C. 1950. *The Psychology of Clothes*. London: Hogarth Press.

Forsythe, S., *et al*. 1985. "Influence of clothing attributes on perception of personal characteristics." In *The Psychology of Fashion*, edited by Solomon, M.R. Lexington, KY: D.C. Heath, pp. 268–277.

Fortunati, L (Ed.). 2002. *Italy, Stereotypes, True and False*. Cambridge, UK: University of Cambridge Press.

Fortunati, L. 2003a. "Mobile phone and the presentation of self." Presented at the Conference *Front Stage/Back Stage: Mobile Communication and the Renegotiation of the Social Sphere*, Grimstad, Norway, June 22–24.

Fortunati, L. 2003b. "The human body: natural and artificial technology." In *Machines That Become Us*, edited by Katz, J.E. New Brunswik, NJ: Transaction Books.

Franzen, A. 2000. "Does the Internet make us lonely?" *European Sociological Review* 16:427–438.

Frønes, I., and Brusdal, R. 2000. *På sporet av den nye tid: Kulturelle varsler for en nær fremtid*. Bergen, Norway: Fagbokforlaget.

Gamson, W. 1975. *The Strategy of Social Protest*. Chicago: Dorsey.

Garfinkel, S.L. 2003. "Leaderless resistance today." *First monday* 3.

Garfinkle, H. 1967. *Studies in Ethnomethodology*. New York: Basic Books.

Gaver, W.W. 1991. "Technology affordances." In *CHI '91*. April 28–May 2, New Orleans, pp. 79–84.

Geertz, C. 1972. "Linguistic etiquette." In *Readings in the Sociology of Language*, edited by Fishman, J.A. The Hague: Mouton, pp. 282–295.

Gergen, K. 2003. "Self and community and the new floating worlds." In *Mobile Democracy: Essays on Society, Self and Politics*, edited by Nyri, K. Vienna: Passagen Verlag, pp. 61–69.

Gibsen, J.J. 1979. *The Ecological Approach to Visual Perception*. New York: Houghton Mifflin.

Giddens, A. 1984. The Constitution of Society: Outline of the Theory of Structuration. Berkeley: University of California Press.

Gillespie, A. 1992. "Communications technologies and the future of the city." In *Sustainable Development and Urban Form*, edited by Breheny, M.J. London: Pion.

Giordano, P.C. 1995. "The wider circle of friends in adolescence." *American Journal of Sociology* 101(3):661–697.

Giuffre, P.A., and Williams, C.L. 1994. "Boundary lines: Labeling sexual harassment in restaurants." *Gender and Society* 8(3):378–401.

Glaser, B., and Strauss, A. 1967. *The Discovery of Grounded Theory*. Chicago: Aldine.

Glaser, B., and Strauss, A. 1971. *Status Passage*. London: Routledge and Kegan Paul.

Godø, H. 2001. "Mobilkommunikasjon, smalbåndsrevolusjonen og telegrafiens renessanse — Fremveksten av en ny type sivilt samfunn." Oslo: Norsk institutt for studier of forskning og utdanning. U-notat 2/2001.

Goffman, E. 1959. *The Presentation of Self in Everyday Life*. New York: Doubleday Anchor Books.

Goffman, E. 1963. *Behavior in Public Places: Notes on the Social Organization of Gatherings*. New York: Free Press.

Goffman, E. 1967. *Ritual Interaction: Essays on Face-to-Face Behavior*. New York: Pantheon Books.

Goffman, E. 1971. *Relations in Public: Microstudies of the Public Order*. New York: Harper.

Gow, G. 2002. "Territorial boundary crossings and regulatory crossroads: The case of wireless enhanced emergency (911) services in Canada." In *Third Wireless World Conference*. 17–18 July 2002, University of Surrey: Digital World Research Centre.

Grahm, S., and Marvin, S. 1996. *Telecommunications and the City: Electronic Spaces and Urban Places*. London: Routledge and Kegan Paul.

Green, N. 2003. "Community redefined: Privacy and accountability." In *Mobile Communication: Essays on Cognition and Community*, edited by Nyiri, K. Vienna: Passagen Verlag, pp. 43–56.

Green, N., and Smith, S. 2002. "'A spy in your pocket'? Monitoring and regulation in mobile technologies." In *Third Wireless World Conference*, 17–18 July 2002, University of Surrey: Digital World Research Centre.

Greenbaum, P., and Rosenfeld, H.M. 1978. "Patterns of avoidance in response to interpersonal staring and proximity: Effects of bystanders on drivers at a traffic intersection." *Personality and Social Psychology* 36(6):575–587.

Grimstveit, L., and Myhre, H. 1995. "A history of mobile communications in Norway." *Telektronikk* 91(4):15–20.

Grinter, R., and Eldridge, M. 2001. "y do tngrs luv 2 txt msg?" In *Proceedings of the Seventh European Conference on Computer-Supported Cooperative Work ECSCW '01*, edited by Prinz, W., *et al.* Dordecht, Netherlands: Kluwer, pp. 219–238.

Groupware Technology. *Information & Computer Science.* Irvine, CA: University of California, Irvine.

Gullestad, M. 1984. *Kitchen-Table Society.* Oslo: Universitetsforlaget.

Gullestad, M. 1992. *The Art of Social Relations: Essays on Culture, Social Action and Everyday Life in Modern Norway.* Oslo: Universitetetsforlaget.

Haavio-Mannila, E., and Snicker, R. 1980. "Traditional sex norms and the innovative function of afternoon dancing." *Research in the Interweave of Social Roles* 1:221–245.

Håberg, S. 1997. "Tilgjengelighet til glede og besvær: En studie av bruk av holdninger til mobiltelefon som ny teknologi." *Institutt for Kulturstudier.* Oslo: Universitetet i Oslo.

Haddon, L. (Ed.). 1997. *Communications on the Move: The Experience of Mobile Telephony in the 1990s.* Farsta: Telia.

Haddon, L. 2000. "The social consequences of mobile telephony: Framing questions." In *Sosiale Konsekvenser av Mobiletelefoni: Proceedings Fra et Seminar om Samfunn, Barn og Mobile Telefoni*, edited by Ling, R., and Thrane, K. Kjeller: Telenor FoU, pp. 2–7.

Haddon, L. 2001. "Domestication and mobile telephony." In *Machines That Become Us*, edited by Katz, New Brunswick, NJ: Transaction.

Haddon, L., and Silverstone, R. 1993. "Teleworking in the 1990s — A view from the home." University of Sussex, Centre for Information and Communication Technologies.

Hägerstrand, T. 1982. "The impact of social organization and environment upon the time-use of individuals and households." In *Internal Structure of the City: Readings from Urban Form, Growth and Policy*, edited by Bourne, L.S. New York: Oxford University Press, pp. 118–123.

Hall, E.T. 1973. *The Silent Language.* Garden City, NY.: Anchor Books.

Hall, E.T. 1989. *The Dance of Life: The Other Dimension of Time.* New York: Anchor Books.

Hård af Segerstaad, Y. 2003a. Dissertation. Dept. of Linguistics, Göteborg University, Göteborg.

Hård af Segerstaad, Y. 2003b. "Language use in Swedish mobile text messaging." Presented at the Conference *Front Stage/Back Stage: Mobile Communication and the Renegotiation of the Social Sphere*, Grimstad, Norway, 22–24 June.

Hardin, G. 1968. "The tragedy of the commons." *Science* 162:1243–1248.

Harp, S.S., *et al.* 1985. "The influences of apparel in responses to television news anchorwomen." In *The Psychology of Fashion*, edited by Solomon, M.R. Lexington, KY: D.C. Heath, pp. 279–289.

Harper, R. 2003. "Are mobiles good or bad for society?" In *Mobile Democracy: Essays on Society, Self and Politics*, edited by Nyiri, K. Vienna: Passagen Verlag, pp. 71–94.

Harter, S. 1990. "Self and identity development." In *At the Threshold: The Developing Adolescent*, edited by Feldman, S.S., and Elliott, G.R. Cambridge, MA: Harvard University Press, pp. 352–387.

Hashimoto, Y. 2002. "The spread of cellular phones and their influence on young people in Japan." In *The Social and Cultural Impact/Meaning of Mobile Communication*, edited by Kim, S.D. 13–15 July, Chunchon, Korea: School of Communication, Hallym University, pp. 101–112.

Haugen, I. 1983. "Tilgjengelighet og sosiale selv." Oslo: Institutt for samfunnsforskning.

Hayden, D. 1984. *Redesigning the American Dream: The Future of Housing, Work and Family Life*. New York: Norton.

Herring, S. (Ed.). 1996. *Computer-Mediated Communication: Linguistic, Social and Cross-Cultural Perspectives*. Amsterdam: John Benjamins.

Herring, S. 2001. "Gender and power in online communication." Bloomington, IN: Indiana University, Center for Social Linguistics. WP-01-05.

Hjorthol, R.J. 2000. "Same city — different options: An analysis of the work trips of married couples in the metropolitan area of Oslo." *Journal of Transportation Geography* 8:213–220.

Hogan, D.P. 1985. "Parental influences on the timing of early life transitions." *Current Perspectives on Aging and Lifecycle* 1:1–59.

Holmes, J. 1981. "Hello–Goodbye: An analysis of children's telephone conversations." *Semiotica* 37(1/2):91–107.

Hutchby, I. 2001. *Conversation and Technology*. Cambridge, MA: Polity.

ICBC. 2001. "The impact of auditory tasks (as in hand-free cell phone use) on driving task performance." ICBC Transportation safety research.

Ito, M. 2003. "Mobile phones, Japanese youth and the replacement of the social contact." Presented at *Front Stage-Back Stage: Mobile Communication and the Renegotiation of the Public Sphere*, Grimstad, Norway, 22–24 June.

ITU. 2003. "Mobile cellular, subscribers per 100 people." http://www.itu.int/ITUD/ict/statistics/ 13 February 2003.

Jackson, T. 1952. "Some variables in role conflict analysis." *Social Forces* 30:323–327.

Jacobs, J. 1961. *The Death and Life of Great American Cities*. New York: Vintage Books.

Johnsen, T.E. 2000. "Ring meg! En studie av ungdom og mobiltelefoni." Department of Ethnology, University of Oslo.

Johnstone, A., *et al.* 1995. "There was a long pause: Influencing turn-taking behavior in human–human and human–computer spoken dialogues." *International Journal of Human Computer Studies* 41:383–411.

Julsrud, T. 2003. "The mobile workplace: How does it influence on behaviour?" Presented at *Front Stage/Back Stage: Mobile communication and the renegotiation of the social sphere*, edited by Ling, R., and Pedersen, Grismstad, Norway, 21–23 June.

Kahlert, H., *et al.* 1986. *Wristwatches: History of a Century's Development*. West Chester, PA: Schiffer.

Kasesniemi, E-L., and Rautiainen, P. 2002. "Mobile culture of children and teenagers in Finland." In *Perpetual Contact: Mobile Communication, Private Talk, Public Performance*, edited by Katz, J. E., and Aakhus, M. Cambridge, UK: Cambridge University Press, pp. 170–192.

Katz, J. E. 1999. *Connections: Social and Cultural Studies of the Telephone in American Life*. New Brunswick, NJ: Transaction Books.

Katz, J.E., and Rice, R.E. 2002. *Social Consequences of Internet Use*. Boston: MIT Press.

Katz, J. E., *et al.* 2001. "The Internet, 1995–2000 Access, civil involvement , and social interaction." *American Behavioral Scientist* 45(3):405–419.

Kavanaugh, A.L., and Patterson, S.J. 2001. "The impact of community computer networks on social capital and community involvement." *American Behavioral Scientist* 45(3):496–509.

Kellner, S. 1977. "Telephone in new (and old) communities." In *The Social Impact of the Telephone*, edited by de Sola Pool, I. Cambridge, MA: MIT Press, pp. 281–298.

Kendon, A. 1967. "Some functions of gaze-direction in social interaction." *Acta Psychologica* 26:26–63.

Klamer, L., *et al.* 2000. "The qualitative analysis of ICTs and mobility, time stress and social networking." Heidelberg: EURESCOM.

Kraut, R., *et al.* 1998. "Internet paradox: A social technology that reduces social involvement and psychological well-being?" *American Psychologist* 53(9):1017–1031.

Krebs, V.E. 2001. "Uncloaking terrorist networks." *Connections* 24(3):

Krogh, H. 1990. "We meet only to part." Doctoral dissertation. Sociology Department, University of Colorado, Boulder.

Kumar, K. 1995. *From Postindustrial to Postmodern Society: New Theories of the Contemporary World.* Oxford, UK: Blackwell.

LaFrance, M., and Mayo, C. 1978. *Moving Bodies: Nonverbal Communication and Social Relationships.* Monterey, CA: Brooks/Cole.

Landes, D.S. 1983. *Revolution in Time: Clocks and the Making of the Modern World.* Cambridge, MA: Belknap Press.

Lange, K. 1993. "Some concerns about the future of mobile communications in residential markets." In *Telecommunication: Limits to Deregulation*, edited by Christofferson, M. Amsterdam: IOS Press, pp. 197–210.

LaRose, R. (Ed.). 1998. *Understanding Personal Telephone Behavior.* Stamford, CT: Ablex.

Lea, M., and Spears, R. 1995. "Love at first byte?: Building personal relationships over computer networks." In *Understudies Relationships: Off the Beaten Track*, edited by Wood, J.T. and Duck, S. Thousand Oaks, CA: Sage.

Leffler, A., *et al.* 1982. "Effects of status differentiation on nonverbal behavior." *Social Psychology Quarterly* 45(3):153–161.

Lemish, D., and Cohen, A.A. 2003. "Tell me about your mobile and I'll tell you who you are: Israelis talk about themselves." Presented at *Front Stage/Back Stage: Mobile Communication and the Renegotiation of the Social Sphere,* Grimstad, Norway, 22–24 June.

Levine, R.V. 1998. *A Geography of Time: The Temporal Misadventures of a Social Psychologist, or How Every Culture Keeps Time Just a Little Bit Differently.* New York: Basic Books.

Licoppe, C., and Smoreda, Z. 2003. "Rhythms and ties: towards a pragmatics of technologically-mediated sociability." Presented at Domestic Impact of Information and Communication Technologies, 5–8 June. Estes Park, Colorado.

Lien, I.L., and Haaland, T. 1998. "Vold og gjengatferd: En pilotstudie av et ungdomsmiljø." Oslo: NIBR.

Lienhard, J.H. 2002. "Engines of out ingenuity." http://www.uh.edu/engines/epi538.htm. 8 December 2002.

Lindmark, S. 2002. "Evolution of techno-economic systems: An investigation of the history of mobile communications." Doctoral Dissertation. Department of Industrial Management, Economics, Chalmers University of Technology, Gothenberg, Sweden.

Ling, R. 1997. "'One can talk about common manners!': the use of mobile telephones in inappropriate situations." In Themes in mobile telephony Final Report of the COST 248 Home and Work group, edited by Haddon, L. Stockholm: Telia.

Ling, R. 1998a. "'It rings all the time': The use of the telephone by Norwegian adolescents." Kjeller: Telenor R&D. Report 17/98.

Ling, R. 1998b. "'She calls, [but] it's for both of us, you know': The use of traditional fixed and mobile telephony for social networking among Norwegian parents." Kjeller: Telenor R&D. Report 33/98.

Ling, R. 1998c. "The technical definition of social boundaries: Video telephony and the constitution of group membership." In *Blurring Boundaries: When Are Information and Communication Technologies Coming Home?* edited by Kant, A., and Mante Meier, E. Stockholm: Telia, pp. 107–130.

Ling, R. 1999. "Traditional fixed and mobile telephony for social networking among Norwegian parents." In *The 17th International Symposium on Human Factors in Telecommunication*, edited by Elstrøm, L. Copenhagen: TeleDanmark.

Ling, R. 2000a. "Direct and mediated interaction in the maintenance of social relationships." In *Home Informatics and Telematics: Information, Technology and Society*, edited by Sloane, A., and van Rijn, F. Boston: Kluwer, pp. 61–86.

Ling, R. 2000b. "The impact of the mobile telephone on four established social institutions." In *ISSEI2000 Conference of the International Society for the Study of European Ideas*. 14–18 August 2000, Bergen, Norway.

Ling, R. 2000c. "We will be reached": The use of mobile telephony among Norwegian youth." *Information Technology and People* 13(2):102–120.

Ling, R. 2001a. "Domestication and mobile telephony." In *Machines That Become Us*, edited by Katz, J.E. New Brunswick, NJ: Rutgers University Press.

Ling, R. 2001b. "It is 'in.' It doesn't matter if you need it or not, just that you have it: Fashion and the domestication of the mobile telephone among teens in Norway." In *Il Corpo Umano tra Tecnologie, Comunicazione e Moda" (The Human Body Between Technologies, Communication and Fashion)*, edited by Fortunati, L. Triennale di Milano, Milano.

Ling, R. 2002. "Preliminary findings from the "Youngster" project: Context-sensitive mobile services for teens." In *Business Models for Innovative Mobile Services*. 15–16 November 2002: Delft University of Technology.

Ling, R. 2003. "The sociolinguistics of SMS: An analysis of SMS use by a random sample of Norwegians." Presented at *Front Stage/Back Stage: Mobile Communication and the Renegotiation of the Social Sphere*, Grimstad, Norway, 22–24 June.

Ling, R. forthcoming. "Mobile communications vis-à-vis teen emancipation, peer group integration and deviance." In *The Inside Text: Social perspectives on SMS in the mobile age*, edited by R. Harper, A. Taylor, and L. Palen. London: Klewer.

Ling, R., and Haddon, L. 2001. "Mobile telephony and the coordination of mobility in everyday life." In *Machines That Become Use*. 18–19 April 2001, Rutgers University, New Brunswick, NJ.

Ling, R., and Hareland, M. 1997. "The cost of being social: User expectations of metering unit displays and cordless telephones." In *Proceedings of the 16th International Symposium on Human Factors in Telecommunication*, edited by Nordby, K. 12–16 May 1997, Oslo, Norway, pp. 365–379.

Ling, R., and Sollund, A. 2002. "Deliverable 13: Planning, operation and evaluation of the field trials." EU IST program. IST-2000-25034 Youngster.

Ling, R., and Yttri, B. 2002. "Hypercoordination via mobile phones in Norway." In *Perpetual Contact: Mobile Communication, Private Talk, Public Performance*, edited by Katz, J.E. and Aakhus, M. Cambridge, UK: Cambridge University Press, pp. 139–169.

Ling, R., and Yttri, B. 2003. "Kontroll, frigjøring og status: Mobiltelefon og maktforhold i familier og ungdomsgrupper." In *På Terskelen: Makt, Mening og Motstand Blant Unge*, edited by Engelstad, F., and Ødegård, G. Oslo: Gyldendal Akademisk.

Ling, R., *et al.* 2001. "The understanding and use of the Internet and the mobile telephone among contemporary Europeans." In *ICUST 2001*. May 2001, Paris.

Ling, R., *et al.* 2002. "E-living deliverable 6. Family, gender and youth: Wave one analysis." IST. Deliverable 6.

Ling, R., *et al.* 2003. "Mobile communication and social capital in Europe." In *Mobile Democracy: Essays on Society, Self and Politics*, edited by Nyri, K. Vienna: Passagen Verlag.

Lippman, A. 1967. "Chairs as territory." *New Society* 20:564–566.

Lohan, E.M. 1997. "No parents allowed!: Telecoms in the individualist household." In *Blurring Boundaries: When Are Information and Communication Technologies Coming Home?* edited by Kant, A. and Mante-Meyer, E., Stockholm: Telia, pp. 131–144.

Love, S. 2001. "Space invaders: Do mobile phone conversations invade peoples' personal space?" In *Human Factors in Telecommunications*, edited by Nordby, K. Bergen, Norway: HFT.

Love, S. 2003. "Does Personality Affect Peoples' Attitude Towards Mobile Phone Use in Public Places?" in *Front Stage/Back Stage: Mobile Communication and the Renegotiation of the Social Sphere*, edited by R. Ling and P. Pedersen. Grimstad, Norway.

Lynd, R. S., and Lynd, H.M. 1929. *Middletown: A Study in Modern American Culture.* New York: Harcourt Brace.

Lynne, A. 2000. "Nyansens makt — en studie av ungdom, identitet og klær." Lysaker: Statens institutt for forbruksforskning. Rapport 4–2000.

Manceron, V. 1997. "Get connected!: Social uses of the telephone and modes of interaction in a peer group of young Parisians." In *Blurring Boundaries: When Are Information and Communication Technologies Coming Home?* edited by Kant, A., and Mante-Meijer, E. Stockholm: Telia, pp. 171–182.

Manning, P.K. 1996. "Information technology in the police context: The 'sailor' phone." *Information Systems Research* 7(1):52–62.

Mante-Meijer, E., and van de Loo, H. 1998. "Blurring of the life spheres: Flexibility and teleworking." In *Blurring Boundaries: When Are Information and Communication Technologies Coming Home?* edited by Kant, A., and Mante-Meijer, E. Stockholm: Telia.

Mante-Meijer, E., *et al.* 2001. "Checking it out with the people — ICT markets and users in Europe." Heidelberg: EURESCOM.

Mars, G., and Nicod, M. 1984. *The World of Waiters.* London: George Allen & Unwin.

Martin, M. 1991. *Hello, Central?: Gender, Technology and Culture in the Formation of Telephone Systems.* Montreal: McGill-Queens.

Marvin, C. 1988. *When Old Technologies Were New: Thinking About Electric Communication in the Late Nineteenth Century.* New York: Oxford University Press.

Marx, G. 1988. *Undercover: Police Survellance in America.* Berkeley: University of California Press.

Mayer, M. 1971. "The telephone and the uses of time." In *The Social Impact of the Telephone*, edited by de Sola Pool, I. Cambridge, MA: MIT Press, pp. 225–245.

Mazon, M. 1984. *The Zoot-Suit Riots: The Psychology of Symbolic Annihilation.* Austin: University of Texas Press.

McCracken, M. 1988. *Culture and Consumption.* Bloomington, IN: Indiana University Press.

McDonald, J. 2002. "Internet Guide to Freighter Travel." http://www.geocities.com/freighter-man.geo/etiquitte.html. 6 December 2002.

Mead, G.H. 1925. "The genesis of the self and social control." *International Journal of Ethics* 35:251–277.

Mead, G.H. 1934. *Mind, Self and Society from the Standpoint of a Social Behaviorist.* Chicago: University of Chicago Press.

Meyrowitz, J. 1985. *No Sense of Place: The Impact of Electronic Media on Social Behavior.* New York: Oxford University Press.
Mininni, G. 1985. "The ontogenesis of telephone interaction." *Repertorio fondamentale di terminologia economico-commerciale Ingles-Italiano* 17(2–3):197–197.
Miyata, K., *et al.* 2003. "The mobile-izing Japanese: Connection to the Internet and Webphone in Yamanashi." Presented at *Front Stage/Back Stage: Mobile Communication and the Renegotiation of the Social Sphere*, Grimstad, Norway, 22–24 June.
MORI. 2000. "I just text to say I love you." http://www.mori.com/polls/2000/lycos. 19 Jan. 2002.
Moyal, A. 1989. "The feminine culture of the telephone: People patterns and policy." *Prometheus* 7(1):5–31.
Moyal, A. 1992. "The gendered use of the telephone: An Australian case study." *Media Culture and Society* 14:51–72.
Mumford, L. 1963. *Technics and Civilization.* San Diago: Harvest.
Murtagh, G.M. 2002. "Seeing the "rules": Preliminary observations of action, interaction and mobile phone use." In *Wireless World: Social and Interactional Aspects of the Mobile Age*, edited by Harper, R. London: Springer.
NHSTA. 1997. "An investigation of the safety implications of wireless communication in vehicles." http://www.nhtsa.dot.gov/people/injury/research/wireless/.% March, 2003.
Nie, N.H. 2001. "Sociability, interpersonal relations, and the Internet: Reconciling conflicting findings." *American Behavioral Scientist* 45(3):420–435.
Nordal, K. 2000. "Takt og tone med mobiltelefon: Et kvalitativt studie om folks brug og opfattelrer af mobiltelefoner." *Institutt for sosiologi og samfunnsgeografi.* Universitetet i Oslo.
Norman, D. 1990. *The Design of Everyday Things.* New York: Doubleday.
NUA. 2002. "Modern flirts use SMS." http://www.nua.ie/surveys/analysis/weekly_editorial/archives/issue1no212.html. 7 February 2003.
Nurmela, J. 2003. "A "great migration" to the information society?: Patterns of ICT diffusion in Finland in 1996–2002." in The good, the bad and the irrelevant, edited by K-H. Kommonen. Helsinki, Finland: COST 269.
O'Connor, K., and Maher, C. 1982. "Change in the spatial structure of a metropolitan region: Work–residence relationships in Melbourne." In *Internal Structure of the City: Readings from Urban Form, Growth and Policy*, edited by Bourne, L.S. Oxford University Press: New York, pp. 406–421.
Ogburn, W.F. 1950. *Social Change.* New York: Viking Books.
Palen, L. 1998. "Calendars on the New Frontier: Challenges of Groupware Technology: Dissertation Information and Communication Science, University of California, Irvine.
Palen, L. 2002. "Mobile telephony in a connected life." *Communications of the ACM* 45(3):78–82.
Palen, L., *et al.* 2001. "Discovery and integration of mobile communications in everyday life." *Personal and Ubiquitous Computing* 5:109–122.
Paragas, F. 2000. "A case study on the continuum of landline and mobile phone services in the Philippines." Presented at the Conference *The Social and Cultural Impact/Meaning of Mobile Communication Conference*, Chunchon, South Korea, 13–15 July.
Park, W.K. 2003. "Mobile phone addiction: A case study of Korean college students." Presented at *Front Stage/Back Stage: Mobile Communication and the Renegotiation of the Social Sphere*, edited by Ling, R., and Pedersen, P. Grimstad, Norway, 22–24 June.
Parker, S. 1988. "Rituals of gender: A study of etiquette, public symbols and cognition." *American Anthropologist* 90(2):372–384.

Parks, M., and Floyd, K. http://www.ascusc.org/jcmc/vol1/issue4/parks.html. 12 January 2004.

Park, M.R. 1996. "Making friends in cyberspace." *Journal of Computer-Mediated Communication* 1(4).

Parks, M.R., and Roberts, L.D. 1998. "'Making MOOsic': The development of personal relationships online, a comparison to their offline counterparts." *Journal of Social and Personal Relationships* 15(4):517–537.

Pedersen, P. 2002. "The adoption of text messaging services among Norwegian teens: Development and test of an extended adoption model." Bergen: SNF Report 23/02.

Piaget, J. 1948. *The Moral Judgment of the Child*. Glencoe, IL: Free Press.

Plant, S. Not dated. "On the mobile: The effects of mobile telephones on social and individual life." http://www.motorola.com/mot/documents/0,1028,333,00.pdf. 12 January 2004.

Polhemus, T. 1994. *Street Style: From Sidewalk to Catwalk. London*. London: Thames and Hudson.

Portes, A. 1998. "Social capital: Its origins and applications in modern sociology." *Annual Review of Sociology* 24:1–24.

Post, P. 2002. "Peggy Post Etiquette for Today." http://magazines.ivillage.com/goodhousekeeping/experts/peggy/qas/0,12875,284570_291705,00.html. 6 Dec. 2002.

Potts, J. 2000. "Wireless phone calls to 911: Steps toward a more effective system." *Currents* 11(2):4–5.

PT. 2003. "De norske telemarked, først halvår 2003." Oslo: Post and teletilsyn. url http://www.npt.no/pt_internet/venstremeny/publikasjoner/telestatistikk/statistikk2003/telehalvaar2003.pdf 11 January 2004.

Putnam, R. 2000. *Bowling Alone: The Collapse and Revival of American Community*. New York: Touchstone.

Quek, F., *et al.* 2000. "Gesture, speech and gaze cues for discourse segmentation." In *IEEE Conference on Computer Vision and Pattern Recognition*, June 2000, Hilton Head Island, SC, pp. 247–254.

Rakow, L.F. 1988. "Women and the telephone: The gendering of a communications technology." In *Technology and Women's Voices: Keeping in Touch*, edited by Kramarae, C. New York: Routledge, pp. 207–229.

Rakow, L.F. 1992. *Gender on the Line*. Urbana, IL: University of Illinois Press.

Rakow, L.F., and Navarro, V. 1993. "Remote mothering and the parallel shift: Women meet the cellular telephone." *Critical Studies in Mass Communication* 10:144–157.

Rasmussen, T. 2003. "Mobile medier og individualisering." Oslo: Uniiveritetet i Oslo, Institutt for medier og kommunikasjon.

Rautiainen, P., and Kasesniemi, E-L. 2000. "Mobile communication of children and teenagers: Case Finland 1997–2000." In *Sosiale konsekvenser av mobiletelefoni: Proceedings fra et seminar om samfunn, barn og mobile telefoni*, edited by Ling, R., and Thrane, K. Kjeller: Telenor FoU, pp. 15–18.

Recarte, M.A., and Nunes, L.M. 2000. "Effects of verbal and spatial imagery tasks on eye fixations while driving." *Journal of Experimental Psychology: Applied* 6(1):31–43.

Redelmeier, D.A., and Tibshirani, R.J. 1997. "Association between cellular-telephone calls and motor vehicle collisions." *The New England Journal of Medicine* 336(7):453–458.

Rheingold, H. 2002. *Smart Mobs*. Cambridge, MA: Persius.

Rice, R.E., and Katz, J.E. 2003. "The telephone as an instrument of faith, hope, terror and redemption: America, 9–11." *Prometheus* 20(3).

Riedman, P. 2002. "U.S. patiently awaits wireless texting that's soaring overseas." *Advertising Age* 73(15):6.

Rivere, C., and Licoppe, C. 2003. "From voice to text: Continuity and change in the use of mobile phones in France and Japan." Presented at *Front Stage/Back Stage: Mobile Communication and the Renegotiation of the Social Sphere,* Grimstad, Norway, 22–24 June.

Robbins, K., and Turner, M.A. 2002. "United States: Popular, pragmatic and problematic." In *Perpetual Contact: Mobile Communication, Private Talk, Public Performance*, edited by Katz, J.E., and Aakhus, M. Cambridge, UK: Cambridge University Press, pp. 80–93.

Robinson, J.P., and Branchi, S. 1997. "The children's hours." *American Demographics* 12:22–24.

Roessler, P., and Hoeflich, J. 2002. "Mobile written communication, or e-mail on your cellular phone." In *The Social Cultural Impact/Meaning of Mobile Communication*, edited by Kim, S.D. Chunchon, Korea: School of Communication, Hallym University, pp. 133–157.

Rogers, E. 1995. *Diffusion of Innovations.* New York: Free Press.

Rombauer, I., and Becker, M. 1964. *The Joy of Cooking.* New York: Signet Books.

Rosenthal, C. 1985. "Kinkeeping in the familial division of labor." *Journal of Marriage and the Family* 47:965–974.

Rubin, L. 1985. *Just friends: The Role of Friendship in Our Lives.* New York: Harper.

Rucker, M., *et al.* 1985. "Effects of similarity and consistency of style of dress on impression formation." In *The Psychology of Fashion*, edited by Solomon, M.R. Lexington, KY: D.C. Heath, pp. 310–319.

Rutter, D. 1987. *Communicating by Telephone.* Oxford, UK: Pergamon Press.

Sakarya, T. 2002. "Mobile phone new emergency and care systems." In *Workshop on Emergency Telecommunications.* 26–27 February, Sophia Antipolis: ETSI.

Saks, H., *et al.* 1974. "The simplest systematics for the organization of turn-taking for conversations." *Language* 50(4):696–735.

Salomon, I. 1985. "Telecommunications and travel: Substitution or modified mobility." *Journal of Transport, Economics and Policy* 19:219–235.

Sandvin, H.C., *et al.* 2002. "Det norske telemarkedet — første halvår 2002." Oslo: Norwegian post and telecommunications authority.

Sattle, J.W. 1976. "The inexpressive male: Tragedy or sexual politics." *Social Problems* 23:469–477.

Savin-Williams, R.C., and Berndt, T.J. 1990. "Friendship and peer relations." In *At the Threshold: The Developing Adolescent*, edited by Feldman, S.S., and Elliott, G.R. Cambridge, MA: Harvard University Press, pp. 277–307.

Schegloff, E., and Saks, H. 1973. "Opening up closings." *Semiotica* 8(4):289–327.

Schegloff, E.A., *et al.* 1977. "The preference for self-correction in the organization of repair in conversation." *Language* 53(2):361–382.

Schwartz, B. 1977. *Queuing and Waiting: Studies in the Social Organization of Access and Delay.* Chicago: University of Chicago Press.

Schwartz, G., and Merten, D. 1967. "The language of adolescents. An anthropological approach to the youth culture." *American Journal of Sociology* 72:453–468.

Sellen, A.J., and Harper, R. 2002. *The Myth of the Paperless Office.* Cambridge, MA: MIT Press.

Sharp, L. 1952. "Case 5: Steel axes for Stone Age Australians." In *Human Problems in Technological Change: A Case Book*, edited by Spicer, E.H. New York: Russell Sage.

Silfverberg, M., *et al.* 2002. "Predicting text entry speed on mobile phones." In *Proceedings of the ACM Conference on Human Factors in Computing Systems – CHI 2000.* New York: ACM, pp. 9–16.

Silverstone, R. 1994. *Television and Everyday Life.* London: Routledge and Kegan Paul.

Silverstone, R. 1995. "Media, communication, information and the 'revolution' of everyday life." In *Information Superhighways: Multimedia Users and Futures*, edited by Emmott, S.J. London: Academic Press, pp. 61–77.

Silverstone, R., and Haddon, L. 1996. "Design and domestication of information and communication technologies: Technical change and everyday life." In *Communication by Design: The Politics of Information and Communication Technologies*, edited by Silverstone, R., and Mansell, R. Oxford, UK: Oxford University Press.

Silverstone, R., and Hirsch, E. 1992. *Consuming Technologies*. London: Routledge and Kegan Paul.

Silverstone R., *et al.* 1992. "Information and communication technologies and moral economy of the household." In *Consuming Technologies: Media and Information in Domestic Spaces*, edited by Silverstone, R., and Hirsch, E. London: Routledge and Kegan Paul, pp. 15–31.

Simmel, G. 1971. *Georg Simmel: On Individuality and Social Forms*", edited by Levine, D.N. Chicago: University of Chicago Press.

Skog, B., and Jamtøy, A.I. 2002. "Ungdom og SMS." Trondheim: ISS NTNU.

Sletten, M.A. 2000. "Ung i Frogn: Rusmiddelbruk, fritidsmønstre, selvbilde og nettverk." Oslo: NOVA. NOVA R 2000: 12, p. 160.

Smoreda, Z., and Thomas, F. 2001a. "Social networks and residential ICT adoption and use." In *EURESCOM Summit 2001 3G Technologies and Applications*. 12–15 November, Heidelberg: EURESCOM.

Smoreda, Z., and Thomas, F. 2001b. "Use of SMS in Europe." http://www.eurescom.de/~ftp-root/web-deliverables/public/p900-series/P9.../w1sms.htm.

Sobel, D. 1996. *Longitude*. London: Forth Estate.

Sørensen, K.H., and Østby, P. 1995. "Flukten fra Detroit. Bilens integrasjon i det norske samfunnet." Trondheim: Senter for teknologi og samfunn, 24/95.

Sproles, G.P. 1985. "Behavioral science theories of fashion." In *The Psychology of Fashion*, edited by Solomon, M.R. Lexington, KY: D.C. Heath, pp. 55–70.

Standage, T. 1998. *The Victorian Internet*. London: Weidenfeld and Nicolson.

Strayer, D., and Johnston, W.A. 2001. "Driven to distraction: Dual-task studies of simulated driving and conversing on a cellular telephone." *Psychological Science* 12(6):462–466.

Strayer, D., *et al.* 2003. "Cell phone–induced failures of visual attention during simulated driving." *Journal of Experimental Psychology: Applied* 9(1):

Stuedahl, D. 1999. "Virklige fantasier: Kibermedia og Goa Kyberia" in *Netts@mfunn*, edited by Braa, K., *et al.* Oslo: Tano Aschehoug, pp. 219–232.

Sullivan, H.S. 1953. *The Interpersonal Theory of Psychiatry*. New York: Norton.

Tanen, D. 1991. *You Just Don't Understand: Men and Women in Conversation*. London: Virago.

Taylor, A. 2003. "Phone-talk and local forms of subversion." Presented at *Front Stage/Back Stage: Mobile Communication and the Renegotiation of the Social Sphere*, Grimstad, Norway, 22–24 June.

Taylor, A., and Harper, R. 2001. "Talking 'Activity': Young people and mobile phones." Presented at *CHI 2001 Workshop: Mobile Communication: Understanding User, Adoption and Design*, Seattle, WA: 31 March–5 April.

Taylor, A., and Harper, R. forthcoming. "The gift of gab: A design-oriented sociology of young people's use of mobiles." *Journal of Collaborative Computing and Computer-Supported Cooperative Work*.

TDG. 2002. "Grameen Telecom's Village Phone Programme: A Multimedia Case Study Prepared for Canadian International development agency." http://www.telecommons.com/villagephone/finalreport.pdf.

Textually.org. 2003. "Climbers on Alpine ridge rescued by text message." http://www.textually.org/textually/archives/001865.htm#001865. 10 Oct. 2003.

Thorns, D.C. 1972. *Suburbia*. London: MacGibbon and Kee.

Tjaden, P., and Thoennes, N. 1998. "Stalking in America: Findings from the National Violence Against Women Survey." Washington, DC: National Institute of Justice/Centers for Disease Control and Prevention, pp. 1–20.

Townsend, A.M. 2000. "Life in the real — mobile telephones and urban metabolism." *Journal of Urban Technology* 7:85–104.

Treichler, P.A., and Kramarae, C. 1983. "Women's talk in the ivory tower." *Communication Quarterly* 31(2):118–132.

Trosby, F. 2003. Personal communication. 31 January 2003.

Trosby, F. no date. "The history of SMS — from a participant's perspective, an unpublished history of the specification of SMS."

Turner, J.H. 1986. *The Structure of Social Theory*. Chicago: Dorsey.

Vaage, O., 1998. "Mediabruks undersøkelse." Oslo: Stastistics Norway.

Veach, S.R. 1981. "Children's telephone conversations." Ann Arbor, MI: University Microfilms International.

Vestby, G.M. 1996. "Technologies of autonomy? Parenthood in contemporary "modern times." In *Making Technologies Our Own: Domesticating Technology into Everyday Life*, edited by Lie, M., and Sørensen, K.E. Oslo: Scandinavian University Press, pp. 65–90.

Walther, J.B. 1993. "Impression development in computer-mediated interaction." *Western Journal of Communication* 57:381–398.

Watson, O., and Graves, T.D. 1966. "Quantitative research in proxemic behavior." *American Anthropologist* 68:971–985.

Weilenmann, A., and Larsson, C. 2002. "Local use and sharing of mobile telephones." In *Wireless World: Social and Interactional Aspects of the Mobile Age*, edited by Brown, B., *et al.* London: Springer, pp. 92–107.

Wellman, B. 1996. "Are personal communities local? A Dumparian reconsideration." *Social Networks* 18:347–354.

Wellman, B. 1999. *Networks in the Global Village: Life in Contemporary Communities*. Oxford, UK: Westview Press.

Wellman, B., and Tindall, D. 1993. "Reach out and touch some bodies: How social networks connect telephone networks." In *Progress in Communication Sciences*. Vol. 12. Edited by Richards, W., and Barnett, G. Norwood NJ: Ablex, pp. 63–93.

Wrolstad, J. 2002. "For mobile phone users, it's location, location location." http://www.wirelessnewsfactor.com/perl/story/18066.html. 24 March 2003.

Yates, S. 1996. "Oral and written aspects of computer conferencing." In *Computer-Mediated Communication*, edited by Herring, S. Amsterdam: John Benjamins.

Young, M., and Buchwald, A. 1965. *White Gloves and Party Manners*. Philadelphia: Robert B. Luce.

Youniss, J. 1980. *Parents and Peers in Social Development: A Piaget–Sullivan Perspective*. Chicago: University of Chicago Press.

Youniss, J., and Smollar, J. 1985. *Adolescent Relations with Mothers, Fathers and Friends*. Chicago: University of Chicago Press.

Zerubavel, E. 1985. *Hidden Rhythms: Schedules and Calendars in Social Life*. Berkeley: University of California Press.

Index

A

Abbreviations in text messaging, 160–161
Accredited members of telephoner's circle, 132–136
Adolescents. *see* Teens and mobile telephones
Adoption of mobile telephony, 4, 9, 11, 12, 17, 21–34, 35, 40, 55, 63, 65, 69, 72, 81–91, 97, 118, 120–121, 165, 171, 172, 175–176, 213–214
  Africa, 14
  Americas, the, 13, 15
  Asia, 3, 14
  Europe, 9, 12–13, 15–16
  Oceania, 11–13
Advanced Mobile Phone System, 8
AMPS. *see* Advanced Mobile Phone System
Andersen, Ben, 182–183
Appropriation, adoption cycle of domestication, 28–31
Asynchronous text messages, 151–152
Automobile-based telephony, 8

B

Baron, Naomi, 148, 158
Beck, Ulrich, 180, 189, 194
Bijker, Wiebe E., 23
Boulding, Kenneth, 176
Bourdieu, Pierre, 178
Bardeen, John, 169
Blacksburg Electronic Village, 182
Brattain, Walter H., 169
Broadcast radio, 7. *see also* Mobile telephony; Radio-based mobile telephony

C

Cameras and mobile telephones, 166
CDMA. *see* Code-Division Multiple Access
Channel assignment, automatic, 8
Code-Division Multiple Access, 8
Commercial broadcast radio, history, 7
Content of text messages, 154–157
  gendering, 164–165
  vocabulary choices, 157–162
  written *versus* spoken language, 162–164
Coordination, dimensions of. *see* Social coordination
Credentialed members of telephoner's circle, 132–136
Cross-cultural influences, 5–6
Cultural lagging, 170

D

Direct dialing, 8
Disruptive nature of mobile telephony, 4, 125–130
de Gournay, Chanal, 190
de Sola Pool, Ithiel, 70
DoCoMo I-mode system
  basics, 10
  text messaging, 145
  versus WAP, 10
Domestication and Information and Communication Technologies, 26–28
  adoption cycle, 28–31
  versus affordances, 27
  analytical perspective, 32–33

Driving and mobile telephone use, 4, 18, 37, 40, 43–44, 49–54, 71, 203, 205, 207
Durkheim, Emile, 67, 174–175, 191

E

Eavesdropping, 140–142
Ellwood-Clayton. Bella, 84, 160, 219
Embarrassment, 125, 140–142
Emoticons in text messaging, 160–161
Ericsson, WAP emergence, 10
Etiquette of mobile telephony, 126–130
Etiquette of timekeeping, 67–69
ETSI. *see* European Telecommunications Standards Institute
EURESCOM P-903 project, 34
European Telecommunications Standards Institute, GSM Communication development, 9

F

Face-to-face communication, *versus* mobile telephony conversations, 129–130
Farley, Tom, 6–7
Fine, Gary, 85, 89, 96, 104, 126, 210
Fischer, Claude, 2–4, 143
Fortunati, Leopoldina, 51, 103, 123, 220
Frønes, Ivar, 94, 165
Full duplex operations, 8

G

General Packet Radio Service, 10
Global System for Mobile Communication standard, 5
  adoption rates, percentages worldwide, 9
  development, 9
  SMS usage, 145
  SMS usage, in United States, 166
GPRS. *see* General Packet Radio Service
Goffman, Irving, 29, 34, 55, 105–106, 125–126, 128, 132–133, 138, 140, 151
Grameen telephone, Bangladesh, 3
  ownership of telephones, 6
Green, Nicola, 53, 190
Greenwich, England, 66
Growth of text messaging
  groups, 152–154
  individuals, 149–152
  outlook, 165–167
GSM communications standard. *see* Global System for Mobile Communication standard
Gullestad, Marianne, 68, 125, 127, 140

H

Haddon, Leslie, 5–6, 9, 26–28, 30–31, 35, 70, 76, 102, 121, 125, 132, 150, 171, 174, 207
Hardin, Garrett, 19, 193
Harper, Richard, 24, 84, 103, 111, 150–153, 155, 218
Harrison, John, 64–65
Hearing-impaired persons and Short Message System, 72
Heavy normative expectations, 125–130
Herring, Susan, 148, 155, 162
Hjorthol, Randi, 174, 207
Hård af Seigerstaad, Ylva, 157, 218

I

I-mode system (DoCoMo)
  basics, 10
  text messaging, 145
  versus WAP, 10
ICTs. *see* Information and Communication Technologies
Imagination, adoption cycle of domestication, 28–31
Incorporation, adoption cycle of domestication, 28–31
Individualism/social capital, 177–181
  ICTs, ad hoc networks, 187–189
  ICTs, Internet, 181–183
  ICTs, mobile telephones, 183–187
Industrial revolution, 172, 175–176
Information and Communication Technologies (ICTs)
  adoption cycle, 28–31

versus affordances, 27
analytical perspective, 32–33
domestication, 26–28
domestication, adoption cycle, 28–31
domestication, analytical perspective, 32–33
domestication, *versus* affordances, 27
social capital/individualism, 181–189
Iterative coordination. *see* Social coordination
Integrated circuits, 171
International roaming, 9
International Telecommunication Union, GSM system adoption percentage worldwide, 9
Internet
access *versus* mobile telephone subscribers, 15–17
role of ICTs and fostering of social capital/individualism, 181–183
Interpersonal situations and mobile telephony
basics, 130–132
break from local setting, 132–136
initiation of calls, 132–136
management of individuals in local setting, 136–138
reemergence into local setting, 138–140
Intrusive nature of mobile telephony, 4, 125–130
Ito, Mizuko, 147, 150, 153, 190
ITU. *see* International Telecommunication Union

J

Julsrud, T., 201

K

Katz, J. E., 16, 36, 46–48, 70, 175, 182–183

L

Landes, David, 64
Licoppe, Christian, 185, 191–192
Love, Steve, 124, 213
Linguistic development and telephone use, 90–93

M

Marine communications
history of mobile telephony, 6
marine chronometer, 64
Mechanical timekeeping, 63
development, 64–65
etiquette, 67–69
standardization of time, 65–67
time-based *versus* mobile-based coordination, 76–80
Meyrowitz, Joshua, 29
Microcoordination. *see* Social coordination
Midcourse adjustment. *see* Social coordination
MMS. *see* Multimedia messages
Mobile telephony. *see also* Radio-based mobile telephony
conversations *versus* face-to-face communication, 129–130
cross-cultural influences, 5–6
eavesdropping, 140–142
history, 6–11
individualism, 179–181
individualism, and ICTs, 181–189
interpersonal situations, basics, 130–132
interpersonal situations, break from local setting, 132–136
interpersonal situations, initiation of calls, 132–136
interpersonal situations, management of individuals in local setting, 136–138
interpersonal situations, reemergence into local setting, 138–140
intrusive and disruptive nature, 4, 125–130
Osborne, Harold S., prediction, 4
ownership differences, 6
social capital/individualism, 177–181
social capital/individualism, and ICTs, 181–189

Mobile telephony (*Continued*)
social consequences, 5
subscribers worldwide, using GSM system, 9
subscription rates, teens, 112
subscription rates, unevenness across social groups, 15–16
subscription rates, *versus* Internet access, 15–17
subscription rates, worldwide, 11–14
system incompatibility limiting usefulness, 15
Motorola, WAP emergence, 10
Multimedia messaging, 166

N
NAMPS. *see* Narrowband Advanced Mobile Phone System
Narrowband Advanced Mobile Phone System, 8
NMT standard. *see* Nordic Mobile Telephone standard
Nokia, WAP emergence, 10
Nordic Mobile Telephone standard, 5, 8–9
Norman, Don, 24, 61
Norway, 3, 5, 8, 11–17, 19, 33–34, 37–38, 50, 68, 78, 84, 87, 94, 102, 109–110, 112, 116, 145–146, 148–151, 155–156, 160, 163, 186, 197–199, 206, 208–209, 214, 216

O
Objectification, adoption cycle of domestication, 28–31
Osborne, Harold S. and mobile telephony, prediction, 4, 169–171
Ownership of mobile telephones, differences, 6

P
Palen, Leysia, 5, 36, 39, 46, 66, 72, 74, 124, 207, 221
Paragas, Fernando, 147
Pedersen, Per, 184, 187
Personal privacy, 53, 169
Phone.com, WAP emergence, 10
Prime Meridian Conference, 66

R
Radio-based mobile telephony. *see also* Broadcast radio; Mobile telephony
Bangladesh, 3
Cheyenne Wells, Colo., 1–3
integration with traditional switched telephony systems, 7
Rakow, Lana, 43, 63, 74, 92, 165, 206–207
Rheingold, Howard, 70, 73, 77, 147, 187–188, 203, 215, 221

S
Safety and security of mobile telephones
acute situations, in, 38–42
abstraction of security, 42–46
basics, 35–38
chronic situations, in, 38–42
crisis situations, 43–48
extraordinary situations, in, 46–48
individuals requiring help, 38–41
loss of personal privacy, 53
reduction of safety, 48–49
reduction of safety, driving, 49–53
Security and safety of mobile telephones
basics, 35–38
crisis situations, 43–48
individuals requiring help, 38–41
loss of personal privacy, 53
reduction of safety, 48–49
reduction of safety, driving, 49–53
Shockley, William B., 169
Short Message System (SMS)
abbreviations, 160–161, 215–216, 219
dominant service, 6
emoticons, 160, 162
GSM development, 9
hearing-impaired persons, 72
linguistic aspects, 158, 162
remote caregiving, 76
social coordination, 79–80

text messaging, 145–146
text messaging, usage in United States, 166
usage, with GSM standard, 145
Silverstone, Roger, 5, 26, 28, 30–31, 103, 125, 171, 174
Simmel, Georg, 106–108, 210
Smart mobs, 187
SMS. *see* Short Message System (SMS)
Smoreda, Zbigniew, 73, 111, 147, 150, 152, 184, 190, 192
Social capital/individualism, 177–181
ICTs, ad hoc networks, 187–189
ICTs, Internet, 181–183
ICTs, mobile telephones, 183–187
Social consequences of mobile telephony, 5
Social coordination
basics, 61–63
mechanical timekeeping, 63
mechanical timekeeping, development, 64–65
mechanical timekeeping, etiquette, 67–69
mechanical timekeeping, standardization of time, 65–67
microcoordination, immediate adjustments, 70–72
microcoordination, iterative, 72–73
microcoordination, softening schedules, 73–76
time-based *versus* mobile-based, 76–80
Social determinism, 17, 23–26, 175
Social networks, 4, 18, 63, 85, 95, 103, 110–112, 151, 153, 174, 178, 183–184, 189, 190, 205, 212
Society and technology, 21–23, 174–177
adoption of technology, 172–174
affordances, 24–26
affordances, *versus* domestication of ICTs, 27
domestication of ICTs, 26–28
domestication of ICTs, adoption cycle, 28–31
history, 172–174
technical/social determinism, 23–25
Softening of schedules. *see* Social coordination
Sonofon, GSM adoption rates, 9
Sound interference, 128
Subscriptions, 3, 9–14, 114, 118, 120, 201, 211, 219
prepaid or "pay as you go", 9, 11–12, 17, 102, 112–113, 204, 211, 214
postpaid, 113–114, 211
Subscription rates of mobile telephones
versus Internet access, 15–17
teens, 112
unevenness across social groups, 15–16
worldwide, 11–14
worldwide, using GSM system, 9
Suburbs, 62, 173–174, 176, 179
Swarming, 77, 188
Switched telephony systems, 7

T

TACS. *see* Total Access Communication System
Taylor, Alex, 103, 150, 152–153, 155, 213
TDMA. *see* Time-Division Multiple Access
Technical determinism, 17, 23–26, 30, 175
Technology and society, 21–23, 174–177
adoption of technology, 172–174
affordances, 24–26
affordances, *versus* domestication of ICTs, 27
domestication of ICTs, 26–28
domestication of ICTs, adoption cycle, 28–31
history, 172–174
technical/social determinism, 23–25
Teens and mobile telephones
child/adolescent development and adoption of telephony, 86–93
emancipation, 93–97, 119–121
introduction, 83–86
payment for services, sources, 114–118
reasons for adoption, 97–99
reasons for adoption, economic aspects, 112
reasons for adoption, fashion, 105–108

Teens and mobile telephones (*Continued*)
  reasons for adoption, functional uses, 99–100
  reasons for adoption, identity issues, 103, 105, 108, 121
  reasons for adoption, social coordination, 100–103, 184
  reasons for adoption, social networking, 110–112
  reasons for adoption, symbolic meaning, 103–110
TeleDanmark, GSM adoption rates, 9
Text messaging or "texting", 145, 147, 162–163, 166, 214. *see also* Short Message System (SMS)
  content, 154–157
  content, gendering, 164–165
  content, vocabulary choices, 157–162
  content, written *versus* spoken language, 162–164
  DoCoMo I-mode, 145
  dominant service, 6
  growth, and groups, 152–154
  growth, and individuals, 149–152
  growth, outlook, 165–167
  GSM standard, 145
  introduction, 145–149
  SMS, 145–146
  SMS, usage in United States, 166
  versus voice telephony, 150
Time-Division Multiple Access, 8
Timekeeping, mechanical, 63
  development, 64–65
  etiquette, 67–69
  standardization of time, 65–67
  time-based *versus* mobile-based coordination, 76–80
Total Access Communication System (TACS), 9
Tragedy of the Commons, 19, 193
Transistors, 7, 169–171
Transoceanic communications, 6

U

United States and mobile telephony, 4, 6, 13, 16, 23, 34, 38, 40, 46, 58, 63, 65–66, 146, 166, 201–202, 214, 220

V

Virtual walled communities, 189–195
Voice telephony, *versus* text messaging, 150
Veach, S. R., 86, 88–89, 91–93, 130, 132, 150, 203
Virtual walled communities, 102, 184

W

WAP. *see* Wireless Application Protocol
Watham Watch Company, 65
Wellman, Barry, 102, 184
Wireless Application Protocol, 9
  versus DoCoMo I-mode system, 10
  emergence, 10
World Trade Center, 43–48